Moral Distress in the Health Professions

Connie M. Ulrich · Christine Grady
Editors

Moral Distress in the Health Professions

 Springer

Editors
Connie M. Ulrich
Lillian S. Brunner Endowed Chair
University of Pennsylvania School
of Nursing,
Philadelphia, PA, USA

Christine Grady
Department of Bioethics
National Institute of Health Clinical Center
Bethesda, Maryland, USA

Department of Medical Ethics and Health
Policy
Perelman School of Medicine,
University of Pennsylvania School
of Medicine
Philadelphia, PA, USA

ISBN 978-3-319-64625-1   ISBN 978-3-319-64626-8   (eBook)
https://doi.org/10.1007/978-3-319-64626-8

Library of Congress Control Number: 2017964513

This Springer imprint is published by Springer Nature
The registered company is Springer International Publishing AG
The registered company address is: Gewerbestrasse 11, 6330 Cham, Switzerland

Acknowledgments

This book would not have been possible without the assistance of our chapter authors. We thank each one of them for taking time out of their busy schedules to contribute to our understanding of moral distress and deepening our resolve that moral distress is a significant problem across clinical units, organizational institutions, geographic locations, and disciplinary homes. We have learned much from them and their perspectives. We would like to especially thank Dr. Kim Mooney-Doyle for her excellent editorial assistance with the book during her postdoctoral work at the University of Pennsylvania. She was a consistent source of optimism and can-do spirit, keeping us on track and moving us forward. We also thank Robin Hermann for sharing her clinical examples of moral distress and the unique stressors that arise in clinical practice in the nursing care of patients and families. Moreover, without the pioneering work on moral distress of our colleague, Dr. Ann B. Hamric, current and further debate and dialogue on defining, measuring, and strategizing about ways to mitigate moral distress would be limited. We thank Ann for her contribution to empirical bioethics work and her assistance with early thoughts on developing a book on moral distress. Finally, we would like to thank all the healthcare professionals who go to work day in and day out and who do their best for every patient and family member. When the storms of life and unexpected illnesses catch patients and families by surprise, they rely on the expertise, goodness, and compassion of nurses, physicians, and others to help navigate their way. We hope this book brings a broader understanding of moral distress to our healthcare professionals, administrators, and others so that we can begin to move toward interprofessional solutions that respect all members of the healthcare team and provide ways to meet the needs of patients and families.

Contents

Contributors

Dominique Benoit Department of Intensive Care Medicine, Ghent University Hospital, Ghent, Belgium

Nancy Berlinger The Hastings Center, Garrison, NY, USA

Katherine Brown-Saltzman Ethics Center, UCLA Health System, Los Angeles, CA, USA

Bo Van den Bulcke Department of Intensive Care Medicine, Ghent University Hospital, Ghent, Belgium

Alyssa M. Burgart Department of Anesthesiology, Perioperative, and Pain Management, Lucile Packard Children's Hospital at Stanford, Stanford Center for Biomedical Ethics, Stanford University, Stanford, CA, USA

Stephen M. Campbell Department of Philosophy, Bentley University, Waltham, MA, USA

Arthur Caplan Division of Medical Ethics, NYU School of Medicine, New York, NY, USA

Anne J. Davis University of California, San Francisco, American Academy of Nursing "Living Legend" in Nursing, San Francisco, CA, USA

Sheila Davis Partners In Health, Boston, MA, USA

Sophia Fantus Center for Medical Ethics and Health Policy, Baylor College of Medicine, Houston, TX, USA

Joseph J. Fins Division of Medical Ethics, Weill Cornell Medical College, New York, NY, USA

New York Presbyterian Hospital Weill Cornell Medical Center, New York, NY, USA

Consortium for the Advanced Study of Brain Injury, Weill Cornell and Rockefeller University, New York, NY, USA

Solomon Center for Health, Law and Policy, Yale Law School, New Haven, CT, USA

Marsha D. Fowler Department of Ethics, Azusa Pacific University, Azusa, CA, USA

Christine Grady Department of Bioethics, National Institutes of Health, Bethesda, MD, USA

Ann B. Hamric School of Nursing, Virginia Commonwealth University, Richmond, VA, USA

Joan Henriksen Mayo Clinic, Rochester, MN, USA

Renatha Joseph Department of Bioethics and Health Professionalism, Muhimbili University of Health and Allied Sciences, Dar es Salaam, Tanzania

Michelle Joy Hospital of the University of Pennsylvania, Philadelphia, PA, USA

Shaké Ketefian School of Nursing, University of Michigan, Ann Arbor, MI, USA

Katherine Kruse Pediatric Critical Care, Department of Clinical Ethics, Children's Hospitals and Clinics of Minnesota, Minneapolis, MN, USA

An Lievrouw Cancer Centre, Ghent University Hospital, Ghent, Belgium

Margaret Lindsey Mount Nittany Medical Center, PA, USA

Margaret Mei Ling Soon Nursing Service, Tan Tock Seng Hospital, Singapore, Singapore

Kim Mooney-Doyle University of Maryland School of Nursing, Baltimore, MD, USA

Georgina Morley Bioethics University of Bristol, Centre for Ethics in Medicine, Bristol, UK

Critical Care Nurse, Barts Heart Centre, Barts Health NHS Trust, London, UK

Baraka Morris Department of Bioethics and Health Professionalism, Muhimbili University of Health and Allied Sciences, Dar es Salaam, Tanzania

Nursing Department of Management, Muhimbili University of Health and Allied Sciences, Dar es Salaam, Tanzania

Lynn Musto School of Nursing, Trinity Western University, Langley, BC, Canada

Linda L. Olson Nursing Regulation, National Council of State Boards of Nursing (NCSBN), Chicago, IL, USA

Member of American Nurses Association Ethics Advisory Board, Silver Spring, MD, USA

Carol L. Pavlish UCLA School of Nursing, Los Angeles, CA, USA

St. Catherine University, St. Paul, MN, USA

Ruth Piers Department of Geriatric Medicine, Ghent University Hospital, Ghent, Belgium

Subadhra D. Rai School of Health Sciences (Nursing), Nanyang Polytechnic, Singapore, Singapore

Ellen M. Robinson Massachusetts General Hospital, Boston, MA, USA

Patricia Rodney UBC School of Nursing, W. Maurice Young Centre for Applied Ethics, UBC, Vancouver, BC, Canada

Robert Truog Medical Ethics, Anaesthesia, and Pediatrics, Center for Bioethics, Harvard Medical School, Boston, MA, USA

Critical Care Medicine, Boston Children's Hospital, Boston, MA, USA

Institute for Professionalism and Ethical Practice, Boston, MA, USA

Connie M. Ulrich Lillian S. Brunner Endowed Chair, University of Pennsylvania School of Nursing, Philadelphia, PA, USA

Department of Medical Ethics and Health Policy, Perelman School of Medicine, University of Pennsylvania School of Medicine, Philadelphia, PA, USA

Tanya Uritsky Pain Management and Palliative Care, Hospital of the University of Pennsylvania, Philadelphia, PA, USA

Mary K. Walton Department of Nursing, Hospital of the University of Pennsylvania, Philadelphia, PA, USA

Bioethics and Nursing, University of Pennsylvania School of Nursing, Philadelphia, PA, USA

Connie M. Ulrich and Christine Grady

> Crucial ethical issues are involved not just in the great questions of life and death, but also in those clinical decisions that, at first sight, appear to be the simplest and most straightforward. [1]

This book is about moral distress, an increasingly familiar term and a common phenomenon in the daily life of those who work in the healthcare professions. Since its original definition by a philosopher more than 30 years ago, moral distress has been defined and studied by various authors and in various ways. Most understand moral distress to occur when a healthcare professional, as a moral agent, cannot or does not act on his or her moral judgment(s) (or what he or she believes to be right in a particular situation) because of institutional or internal constraints. Addressing moral distress requires attention to the everyday ethical or moral concerns that appear routine but nonetheless can challenge both the patient-clinician relationship and the mental and physical well-being of healthcare providers. Some worry that if the rising tide of moral distress is not addressed, the professional and moral integrity of health professionals is at risk. Addressing moral distress also crucially requires attention to the environments and systems in which healthcare workers care for patients. Some commentators have suggested that it is time to either abandon the negative label of moral distress or to move beyond it to discuss how we can build or promote resilience within constricted working environments. We believe that moral distress is real and here to stay but that it is also sometimes exaggerated or misunderstood. We further believe that compromised integrity is not an inevitable

C.M. Ulrich (✉)
Lillian S. Brunner Endowed Chair,
University of Pennsylvania School of Nursing, Philadelphia, PA, USA

Department of Medical Ethics and Health Policy, Perelman School of Medicine,
University of Pennsylvania School of Medicine, Philadelphia, PA, USA
e-mail: culrich@nursing.upenn.edu

C. Grady (✉)
Department of Bioethics, National Institutes of Health Clinical Center, Bethesda, MD, USA
e-mail: CGrady@cc.nih.gov

© Springer International Publishing AG 2018
C.M. Ulrich, C. Grady (eds.), *Moral Distress in the Health Professions*,
https://doi.org/10.1007/978-3-319-64626-8_1

result of experiencing moral distress and that sensitivity to moral challenges can be a source of growth and learning. This book was motivated by a desire to describe the experiences of moral distress and identify commonalities and differences experienced by many diverse healthcare professionals. The aim is to recognize the seriousness of moral distress beyond the nursing discipline, give voice to a community of health professionals facing challenging and complex ethical issues, and promote ongoing dialogue both within and outside academic and clinical communities on the physical and psychological toll of moral distress and strategies for mitigating it.

Healthcare providers are a great and essential asset when sickness occurs. Patients, and the public in general, rely on the substantive body of knowledge and clinical expertise of a team of clinicians dedicated to the care and well-being of patients within designated hospital settings or medical centers. This team includes nurses, physicians, social workers, pharmacists, psychiatrists, researchers, students in training, and many others. This talented group of professionals cares for the young, the old, and the sickest of the sick, often with little fanfare or recognition. Every day these clinicians assess, interpret, analyze, and interact with human health and illness. While caring for patients and their families, healthcare professionals share and reflect on the joys and sorrows that accompany these interactions. And in many ways, they are suffering too.

Today, healthcare clinicians face mounting pressures as they care for their patients, pressures and resulting stress that can impact their own health and well-being. Evidence suggests that clinicians commonly suffer from physical and psychological health-related problems, including but not limited to depression, burnout, fatigue, and moral distress [2–6]. Everyday work within healthcare systems is inherently ethical; clinicians seek to help patients and their families understand complicated diagnoses or procedures, weigh different treatment options and balance medication regimens, and assist with emotionally charged decisions at every stage of life. Healthcare professionals strive to use their knowledge, skills, and expert training to act in the best interests of their patients and families and, at the same time, uphold their professional and ethical values, norms, and principles. But, patient care within the clinical arena is increasingly complex as different groups of stakeholders may have divergent goals, goals that at times seem incompatible, creating unease, ethical tensions, and conflict. The pressures healthcare clinicians face stem not only from the variable and changing demographics and needs of patients and competing demands for often limited resources but also from individual stressors and organizational workplace issues they encounter on a daily basis. Healthcare clinicians often work long hours in fast-paced and complex environments where they are engaged in thorny ethical issues that might seem intractable.

Pavlish and colleagues attribute ethical conflicts to "extended life spans, increased technology, the public's unrealistic expectations of medical care, greater cultural and religious diversity, more emphasis on patient's rights, shifts in healthcare financing, and limited resources" [7]. Indeed, our global society is aging, and healthcare professionals are encountering more racially, ethnically, and culturally diverse patients requiring skilled communication in addressing the preferences and goals of patients and their families. The United States, for example, expects that more than 20% of its population will be 65 years old or older by 2030 [8]. And by 2050, 1.6 billion individuals worldwide will be in this age cohort [9]. Of course, although "people are living longer, that does not necessarily mean that they are living healthier" [9]. In fact, chronic illness is significant in older populations; and many older adults spend time in intensive care environments at the end of life with costly interventions that provide little benefit. Despite the reality as stated by Atul Gawande that "dying and death confront every new doctor or nurse" [10], clinicians continue to struggle to guide patients and families with end-of-life decisions, especially when they do not agree with the decisions made. Everyone who intimately cares for patients at the end of life can recognize the glimmer of hope when a laboratory value shows stability or the next treatment improves physiological functioning but can also recognize the dread and sadness when it does not [11]. Initiating difficult conversations regarding transitions in care from curative to palliative can be challenging and stressful for all involved. Studies suggest that the lack of educational preparation on advance care and end-of-life planning for both nurses and physicians hampers their ability to confidently discuss end-of-life concerns with seriously ill patients and their families [12–14].

Ethical issues and conflicts are commonplace in today's healthcare environment. Arthur Caplan, a prominent bioethicist, noted that the "ordinariness" of the day-to-day ethical questions and problems within health institutions often "appears mundane or banal" (p. 38) [15]. Although he was directly speaking about medical residents in nursing homes at the time, the same can be said about nurses, physicians, pharmacists, social workers, and other healthcare providers who spend many hours in the service of caring for patients and families within hospitals and other acute, long-term care and community settings. Day-to-day work-related ethical questions are part of the operational life within these walls and include but are not limited to: Do I have enough staff to safely care for these patients? Should I report a near miss that didn't harm the patient? Who is going to share the news with my patient that she has advanced cancer for which chemotherapy and other treatments will likely be of little benefit? And what if she insists on aggressive treatment that has little hope? How should I discharge my patient when there is no assistance at home, but the patient's insurance dictates the number of hospital days covered? What should I do when my patient refuses his medication and has become agitated? Should I tell the family of a patient with Alzheimer's who have asked my opinion twice about long-term care options what I would do if I were in their position? Should I call the attending physician about a questionable laboratory value and risk potential backlash at 3 am in the morning? Do I suggest that my patient, who doesn't understand the procedure being proposed and is already scheduled in the operating room, withhold his signature from the informed consent document until the clinical fellow has more time to explain it in detail? How do I help my patient understand the risks and trade-offs of treatment choices? Every day ethical questions arise in the patient-clinician relationship that, as Caplan notes, might seem at odds with how one thinks of ethical or moral questions; but "ethics concerns not only questions of life and death but how one ought to live with and interact with others on a daily basis. The ethics of the ordinary is just as much a part of health care ethics as the ethics of the extraordinary" (p. 38) [15]. Ulrich and colleagues found that clinicians might question or have concerns about what is morally right or wrong in many commonly encountered clinical situations [5]. Ordinary questions—along with the extraordinary—can challenge professional and ethical practice on a daily basis.

Nurses, for example, identify a host of ethical challenges in clinical care (and research). They often report participation in morally distressing situations, feeling powerless to change or alter the course of certain decisions, and consequently experiencing psychological, physical, and moral stress. The example below is an illustration of moral distress that occurs when healthcare clinicians are placed in a morally undesirable position and suffer from self-directed negative emotions such as anger, guilt, and remorse.

> Mr. Roberts is a terminally ill patient with a diagnosis of end-stage lung disease in the intensive care unit. Prior to meeting with Mr. Roberts, the nurse and attending physician met with his family to discuss their concerns. During this meeting, the family insisted that the patient not be told about hospice options or other comfort care options because if he knew his prognosis, "he would just give up and die." In a previous conversation earlier that day with Mr. Roberts, however, the nurse caring for him learned that he was in fact aware that he would not receive the lifesaving transplant that he needed and that he knew that he was going to die. The patient was also told by the physician that there were no treatment options for him and that he wouldn't be able to go home as he wished based on his current level of care needs. The patient was competent, awake, and seemed very aware of what was happening with his care. He was also on high levels of oxygen which made it unsafe for him to eat or drink, although Mr. Roberts continued to ask for food and drink and became increasingly angry when these were denied. The nurse was also worried that any food or drink could cause the patient to arrest, and according to current orders, he remained a full code. Without telling the patient that he could die without the oxygen, the primary nurse for Mr. Roberts explained the risks as much as she could. She also spoke with the family about allowing Mr. Roberts to eat and drink but that he would need to be fully informed of the risks. Again, the family refused, indicating that they did not want him to know all the information surrounding his prognosis. The nurse felt as though she was lying to the patient every time she walked into his room to care for him. She became angry at the family for placing her in a position that went against both her personal and professional values even though it was what the family wanted.

Ethical issues are not isolated to those who are on the wards every day as full-time healthcare employees. Medical residents and students also express unique ethical concerns reflective of their positions with healthcare systems. Similar to their nursing colleagues, they are subject to the decision-making authority hierarchy, sandwiched between an attending physician and others on the healthcare team. They often worry about speaking out; at times questioning their own competency to do so, and fearing the potential consequences that might come their way. It is difficult to expect medical and nursing students to lead—and to become future leaders—if they are not led or given the space to identify, critique, and resolve the moral questions they encounter. The increase in violence perpetrated against healthcare clinicians is also troubling. In fact, in one study, 76% of hospital nurses reported patient and visitor aggression and violence over the past year—emotional, physical, or verbal [16]. Being at the bedside for protracted periods of time increases the potential risk of harm by angry and stressed patients or family members, especially for clinicians who work in emergency or psychiatric settings. Continual exposure to these ethical issues can lead to moral distress, emotional and physical exhaustion, injuries, and, ultimately, declines in a quality healthcare workforce.

1.1 The Organization of the Book

The idea of writing a book on moral distress first percolated in 2008 at a conference at the American Society of Bioethics and Humanities. Although it has taken us some time…indeed, almost 10 years…to bring this to fruition, we hope that this book helps all of those who have faced moral distress in their clinical work and enlightens those who have not, recognizing that moral distress is a present-day reflection of the real-time minute-to-minute and hour-to-hour reality of patient care delivery in complex systems. Since that time, there has been an explosion of scholarly discourse, research, and commentary on moral distress demonstrating that many healthcare clinicians experience this phenomenon, regardless of their professional or practice discipline. Indeed, moral distress is not going away any time soon.

This book seeks to challenge readers on the following questions:

1. What do we know about moral distress and how has it been defined since its original definition in 1984? Much of the early conceptual and empirical work on moral distress focused on the nursing profession because of nurses' distinctive position in the healthcare hierarchy. Today, however, we know that many other healthcare disciplines are similarly distressed.
2. What do we know about moral distress from empirical studies, and where are the gaps? Does the research adequately capture the phenomenon across settings and disciplines? We also need more research that moves beyond mere description of the problem to help get us closer to understanding its impact and ways to prevent or mitigate it.
3. How should we understand the value of an ethical climate in the workplace, what does this look like, and how should we promote ethical climates in the healthcare workplace? What role do the values of an organization and its leadership play in enhancing or impeding the ethical care of patients and support of healthcare professionals?
4. What are the lessons we can learn from experiences of moral distress from diverse healthcare disciplines, both domestic and international? Do healthcare clinicians in both the developed and developing worlds share similar yet unique ethical challenges?
5. What are the most significant or important reasons that healthcare professionals might feel trapped and unable to do what they think is right (hence experience moral distress)? And, what does moral success look like within healthcare institutions today?

The complexity of healthcare institutions, the increasing patient acuity levels and the evolution of societal ills (such as the opioid crisis in the United States and the management of substance abuse patients, and Ebola in the developing world), the corporatization of healthcare, and the continual development of technology (e.g., precision science and genomics) to address some of the most recalcitrant health problems, both domestically and internationally, can all contribute to challenges that result in moral distress for healthcare clinicians as they address the ethical issues before them. Organizations have responsibilities to provide safe working environments for their employees; and, to some degree, organizational systems and the organizational culture should identify successes or failures in preventing and resolving morally distressing situations. Learning from our moral mistakes and successes can also be instructive. Building a culture and climate of open discourse without fear of retribution that is responsive to ethical issues and concerns recognized by nurses, physicians, and others supports the overall health and well-being of those responsible for patients and families who place their trust in these clinicians during their most vulnerable moments. It is our hope that this book will lead to further dialogue, research, and conceptual clarification of moral distress that includes both normative and empirical scholarship as an important area of bioethics inquiry and health services research. "What we do in our communities and companies—the public policies we put in place, the ways we help one another—can ensure that fewer people suffer." (p. 11) [17].

References

1. Truog RD, Brown SD, Browning D, Hundert EM, Rider EA, Bell SK, et al. Microethics: the ethics of everyday clinical practice. Hast Cent Rep. 2015;45(1):11–7.
2. Hamric AB, Blackhall LJ. Nurse-physician perspectives on the care of dying patients in intensive care units: collaboration, moral distress, and ethical climate. Crit Care Med. 2007;35(2):422–9.
3. Hamric AB. Moral distress and nurse-physician relationships. Virtual Mentor. 2010;12(1):6–11.
4. Bodenheimer T, Sinsky C. From triple to quadruple aim: care of the patient requires care of the provider. Ann Fam Med. 2014;12(6):573–6.
5. Ulrich CM, Taylor C, Soeken K, O'Donnell P, Farrar A, Danis M, et al. Everyday ethics: ethical issues and stress in nursing practice. J Adv Nurs. 2010;66(11):2510–9.
6. Ulrich C, O'Donnell P, Taylor C, Farrar A, Danis M, Grady C. Ethical climate, ethics stress, and the job satisfaction of nurses and social workers in the United States. Soc Sci Med. 2007;65(8):1708–19.
7. Pavlish C, Brown-Saltzman K, Jakel P, Fine A. The nature of ethical conflicts and the meaning of moral community in oncology practice. Oncol Nurs Forum. 2014;41(2):130–40.
8. Ortman JM, Velkoff VA, Hogan H. An aging nation: the older population in the United States population estimates and projections: current population reports. https://www.census.gov/prod/2014pubs/p25-1140.pdf; 2014.
9. Cire B. World's older population grows dramatically. Retrieved from https://www.nia.nih.gov/news/worlds-older-population-grows-dramatically; 2016.
10. Gawande A. Being mortal: medicine and what matters in the end. 1st ed. New York: Metropolitan Books: Henry Holt & Company; 2014. 282 p.
11. Ulrich C. End of life futility conversations: when language matters. Perspectives in Biology and Medicine. 2017;60(3): 433–7. (in press)

12. Institute of Medicine (U.S.) Committee on Approaching Death: Addressing Key End-of-Life Issues. Dying in America: improving quality and honoring individual preferences near the end of life, vol. xxv. Washington, D.C.: The National Academies Press; 2015. 612 p.
13. Long AC, Downey L, Engelberg RA, Ford DW, Back AL, Curtis JR. Physicians' and nurse practitioners' level of pessimism about end-of-life care during training: does it change over time? J Pain Symptom Manag. 2016;51(5):890–7.e1.
14. Visser M, Deliens L, Houttekier D. Physician-related barriers to communication and patient- and family-centred decision-making towards the end of life in intensive care: a systematic review. Crit Care. 2014;18(6):604, 1–19.
15. Caplan AL. The morality of the mundane: Ethical issues arising in the daily lives of nursing home residents (pp. 37–50). In RA Kane and AL Caplan (Eds). Everyday ethics: resolving dilemmas in nursing home life, vol. xvii. New York: Springer; 1990. 331 p.
16. Speroni KG, Fitch T, Dawson E, Dugan L, Atherton M. Incidence and cost of nurse workplace violence perpetrated by hospital patients or patient visitors. J Emerg Nurs. 2014;40(3):218–28.
17. Sandberg S, Grant AM. Option B: facing adversity, building resilience, and finding joy. 1st ed. New York: Alfred A. Knopf; 2017. 226 p.

Lynn Musto and Patricia Rodney

2.1 Moral Distress: Evolution of the Concept

When theory and practice in healthcare ethics started to evolve in the late 1970s, there emerged a growing consensus about how ethical principles ought to guide healthcare delivery [1, 2], yet the well-being of healthcare providers received relatively little attention. This lack of attention started to change with American philosopher Andrew Jameton's groundbreaking writing about moral distress in his book on nursing ethics [3]. Jameton's book, his subsequent publications, and the early related research by nurse scholars such as Fry, Harvey [4], Hamric [5], and Wilkinson [6] initiated an important conceptual and practical shift. This shift has helped all of us involved in healthcare to recognize that the moral experiences of healthcare providers affect the quality of healthcare delivery and also the well-being of the providers themselves [7–9].

In this chapter, we offer a further contribution to growing contemporary commentaries on how the concept of moral distress has evolved and how it has been applied, including its pitfalls and promises. Our intent is to continue to support what we see as a lively and promising dialogue about moral distress in nursing, other healthcare provider groups, and healthcare ethics in general. On the basis of our experiences in practice and research, it is our conviction that continuing to wrestle with the clarity of the concept, its application, and the implications for *practice* (including leadership) in healthcare remains important. We believe that supporting nurses and all other healthcare providers as moral agents operating in complex

L. Musto (✉)
School of Nursing, Trinity Western University, Langley, BC, Canada
e-mail: Lynn.Musto@twu.ca

P. Rodney
UBC School of Nursing, W. Maurice Young Centre for Applied Ethics, UBC,
Vancouver, BC, Canada

© Springer International Publishing AG 2018
C.M. Ulrich, C. Grady (eds.), *Moral Distress in the Health Professions*,
https://doi.org/10.1007/978-3-319-64626-8_2

organizational structures is prerequisite to offering effective and ethical healthcare and fostering a sustainable healthcare workforce.

We will therefore provide an overview of the evolution of the definition of moral distress, outline some of the critiques of the concept that have shaped our exploration, and point to areas for further research and development. We close our chapter with conceptual and practical recommendations for nursing, other healthcare provider groups, and for the structure and delivery of healthcare. It is important to note that while the study of moral distress was initiated in the United States, it is now also increasingly being addressed by colleagues from diverse parts of the globe—including, for instance, Australia [10, 11], Brazil [12], Canada [13], Ireland [14], and Iran [15]. While we will not be undertaking a full international analysis of the concept of moral distress, we will point to some of the implications of the expanding global interest in the concept toward the end of this chapter.

2.2   Conceptual Origin and Evolution of the Definition

Healthcare ethics evolved in response to the significant values-based challenges that healthcare providers faced in trying to provide competent, effective, and equitable care in the face of decisions regarding the effective deployment of healthcare technology and equitable access to healthcare resources[1] [2, 18]. As we have noted in our introduction to this chapter, attention to the well-being of healthcare providers started to emerge more directly when Andrew Jameton, a philosopher, was working with nurses and observed that "moral and ethical problems in the hospital could be sorted into three different types," moral uncertainty, moral dilemmas, and moral distress [3]. Jameton's original definition of moral distress stated that the experience arose "when one knows the right thing to do, but institutional constraints make it nearly impossible to pursue the right course of action" [3]. In identifying moral distress, Jameton put into words a collective experience that occurred when nurses confronted situations that created a conflict in their professional values—a conflict that often ultimately left the nurse with the sense that they had failed to live up to their moral obligations to the patient.

Although identification of the concept captured the attention of nursing scholars, when nurse researchers and researchers in other disciplines began to operationalize the definition, it soon became clear that there were gaps. As research on moral distress progressed, scholars articulated some of those gaps, including potential

---

[1] It is worth noting that early healthcare ethics work was largely silent on access to resources for *health*, such as ethnicity, education, and income. Although equitable access to resources for health is receiving more attention in contemporary healthcare ethics work (e.g., [16]), much more work is needed. Indeed, Varcoe et al. [17] argue that "…the same socio-political values that tend to individualize and blame people for poor health without regard for social conditions in which health inequities proliferate, hold responsible, individualize and even blame healthcare providers for the problem of moral distress" (p. 52).

conflation of moral distress and psychological or emotional distress, leading to a call for researchers to focus on the *ethical* component of moral distress [14, 19]; the view of moral distress as a *linear* process [20, 21]; the need for a richer understanding of moral distress as a process that unfolds over time [21, 22]; the actual location of *constraints* on moral action, for example, locating *constraints* to action internal to the nurse or externally within the institution [6, 23, 24]; the need to uncouple constraint as a necessary cause of distress and include related experiences such as *conflict* [25]; lack of clarity around what constitutes the right course of action and the role of action in general [21, 24, 26]; as well as a *lack of clarity overall* about the concepts that underpin moral distress [14, 17, 27].

As a result of working with an evolving definition, researchers continue to seek to refine the definition, and our full understanding of the concept remains "under construction" (see, e.g., Fourie [25]). One of the consequences is a growing list of definitions that seek to incorporate our developing understanding of moral distress (see Table 2.1). The table in this chapter is not intended to be exhaustive; rather, the intention is to provide examples that illustrate the evolution of the concept as scholars and researchers incorporate new insights into the definition of moral distress in

Table 2.1 Evolving definitions of moral distress

Authors	Definition
Jameton [3]	Moral distress arises when one knows the right thing to do, but institutional constraints make it nearly impossible to pursue the right course of action
Wilkinson [6]	Psychological disequilibrium and negative feeling state experienced when a person makes a moral decision but does not follow through by performing the moral behavior indicated by that decision
Jameton [26]	*Initial moral distress* involves the feelings of frustration, anger and anxiety people experience when faced with institutional obstacles and conflict with others about values
	Reactive moral distress is the distress people feel when they do not act upon their initial distress
Hanna [28]	Moral distress occurs in the context of situations that have moral implications embedded within them, where the moral end, an inherent rightness or goodness, is understood to exist and understood to be or have been threatened, harmed or violated.
Austin et al. [29]	The state experienced when moral choices and actions are thwarted by constraints
Kälvemark et al. [30]	Traditional negative stress symptoms that occur due to situations that involve ethical dimensions and where the healthcare provider feels she/he is not able to preserve all interests and values at stake
Nathanial [31]	Moral distress is pain affecting the mind, the body, or relationships that results from a patient care situation in which the nurse is aware of a moral problem, acknowledges moral responsibility, and makes a moral judgment about the correct action, yet, as a result of real or perceived constraints, participates, *either by act or omission*, in a manner he or she perceives to be morally wrong

(continued)

Table 2.1 (continued)

Authors	Definition
Mitton et al. [32]	Moral distress is the suffering experienced as a result of situations in which individuals feel morally responsible and have determined the ethically right action to take, yet due to constraints (real or perceived) cannot carry out this action, thus committing a moral offence
Varcoe et al. [17]	The experience of being seriously compromised as a moral agent in practicing in accordance with accepted professional values and standards
Rodney et al. [33]	What nurses (or any moral agents) experience when they are constrained from moving from moral choice to moral action—an experience associated with feelings of anger, frustration, guilt, and powerlessness
Crane et al. [11]	The experience of psychological distress that results from engaging in, or failing to prevent, decisions or behaviors that transgress, or come to transgress, personally held moral or ethical beliefs
Barlem and Ramos [34]	The feeling of powerlessness experienced during power games in the micro-spaces of action, which lead the subject to a chain of events that impels him or her to accept imposed individualities, have his or her resistances reduced and few possibilities of moral action; this obstructs the process of moral deliberation, compromises advocacy and moral sensitivity, which results in ethical, political and advocational inexpressivity and a series of physical, psychical, and behavioral manifestations
Campbell et al. [21]	One or more negative self-directed emotions or attitudes that arise in response to one's perceived involvement in a situation that one perceives to be morally undesirable

an effort to bring further clarity and move the concept forward. Despite this growing list of definitions and the scholarly analyses that have generated them, much of the current research on moral distress continues to be based on the foundation created by the earliest definitions of moral distress offered by Jameton [3] or Wilkinson [6]. Research studies over the years have indicated that causes of moral distress in nursing are "varied, and include conflict with other clinicians, an excessive workload, and challenges with end-of-life decision making" [7].

Tracking the Evolution in Our Understanding of Moral Distress It is important to note that the concept originated from within the discipline of nursing, and as such, the definition and early exploration of the concept have been influenced by the disciplinary culture of nursing. An example of the disciplinary influence on the definition is seen in the discovery that one of the contributing elements to the experience of moral distress in nursing is a lack of decision-making authority in relation to resource allocation or clinical care [10]. Although nurses do, for the most part, have less authority to make decisions in healthcare organizations, physicians also experience moral distress *because* they are responsible for the decisions they make [21,

35–37]. These disparate findings suggest that interpreting research findings through a solitary disciplinary lens may unintentionally limit our interpretations. The predominance of a focus on moral distress in *nursing* is, to a significant extent, "ethnocentric" and does not serve our colleagues in other healthcare provider groups well [7, 9]. It is clear that experiencing moral constraints and/or moral conflicts (however we define them) transcends professional disciplinary boundaries[2] [7, 9]. Research on moral distress as a transdisciplinary experience has added depth and breadth to our understanding of the concept. As indicated above, much of the multidisciplinary work continues to use early definitions of moral distress that are imbued with the nursing perspective on the experience. The significance of understanding that moral distress crosses disciplinary boundaries points to the necessity of moving the definition itself beyond the discipline of nursing to a level that can account for the range of the experiences of moral distress in healthcare.

2.3   Challenges and Critiques of the Definition

As we understand the original definition of moral distress, it was predicated on three main assumptions: (a) that nurses make moral judgments, (b) that they do not act on those moral judgments; and (c) that their inaction is related to institutional constraints [35]. In naming moral distress, Jameton made a distinction between personal and professional values [3]. Hanna [19] provides a critique of this "artificial" separation stating that the consequence would be that "personal values and beliefs that translate into private thoughts and deeds meant that any person's efforts would have no bearing on the social fabric of the community. Yet communities are comprised of the thoughts, words, and deeds of many people" (p. 75). The connection we want to highlight is that the moral obligations of a profession are established in and through community (society) and as such are based on societal values, which are both personal and professional. We will come back to this point when we discuss reciprocity between structures and agents laying the ground for recommendations aimed at developing a greater understanding of, and developing interventions for, moral distress.

Each of the assumptions listed above presents a unique set of challenges that we will summarize. Hanna [19], one of the first nurse scholars to offer a thorough critique of the definition, pointed to the assumption that the nurse had knowledge, and certainty, about what was the right course of action in a given situation.[3] Johnstone and Hutchinson [40] pick up on this critique and push it further by distinguishing between making an ordinary moral judgment based on personal opinion and a moral judgment based on "sound critical reflection and wise reasoning" (p. 4). Johnstone and Hutchinson [40] also draw on findings from the literature in

---

[2] For example, in a piece in *Narrative Inquiry in Bioethics*, a healthcare provider, Cheryl Mack, explores her response to the moral uncertainty she experienced in a complex organ donation situation [38].

[3] For further information on this critique, we refer the readers to McCarthy and Gastmans [39], Johnstone and Hutchinson [40], Hanna [19] and Repenshek [41].

neuroscience and moral psychology that suggest moral judgments are based in intuition and that people use post hoc justification to support their moral judgments. Further, the authors assert that nurses' judgments are grounded in personal, rather than professional, values [40]. From our perspective, these critiques are examples of how development of the concept has been influenced by an ethnocentric perspective based in nursing. By this, we mean that similar critiques could be leveled at all disciplines, not just nurses. However, as a number of scholars have noted [14, 40, 42], because moral distress came out of the nursing discipline, there may be a historical conflation of the concept with disciplinary narratives, such as moral suffering and powerlessness. We therefore believe that in order to develop conceptual clarity on the assumptions that underpin the definition, it is imperative to move the concept beyond one single discipline. Further, scholars from outside of nursing, such as philosophy and medicine, have begun to question the role of moral *uncertainty* in the experience of moral distress [21, 38], thereby extending our understanding of moral distress beyond an assumption of moral certainty to a place of engaging with moral ambiguity. It is also not clear that one can easily distinguish personal from professional values in making moral judgments [19] without greater comprehension of how moral judgments are actually made. Overall, these critiques highlight the need to draw from insights across academic disciplines—for example, philosophy, bioethics, and moral psychology[4]—in order to continue work to develop a comprehensive understanding of moral distress for nurses as well as other healthcare providers.

The role of *action*, or the enactment of moral agency, has been gaining attention in the literature on moral distress as researchers have been encouraged to seek conceptual clarity. In several of the definitions listed in Table 2.1, the language used to describe moral action sets up a binary; individuals either take action or they do not take action. Jameton's original definition suggested a linear conception of moral distress with action as the fulcrum.[5] The assumption was that if the nurse, or other healthcare provider, took action, they would not experience moral distress [27]. Applying a more nuanced lens to action revealed that nurses, and other healthcare providers, frequently do take action; however, their actions are often not recognized [24, 43]. Other research suggests that taking action not only does not alleviate the experience, it may also *contribute* to moral distress [43–46]. In a study that examined nurses' responses to morally distressing situations, Varcoe and Pauly [43] identified both the extensive actions taken and the ways in which these actions were dismissed within the healthcare system. These authors highlight the questionability of having the phrase "unable to act" as one of the assumptions that unpins the definition of moral distress and instead encourage examination of continuous actions that may fail to resolve the distressing situation. This perspective of action has contrib-

---

[4] For example, social psychologist Bandura's writing about *moral disengagement* can help us to understand how healthcare providers may respond to moral distress [33].

[5] We believe that Jameton's understanding of action was more nuanced than his definition suggests and refer readers to his 1993 article *Dilemmas of moral distress: Moral responsibility and moral practice* for a more in-depth view of his perspective on action.

uted to a view of moral distress as a relational experience where moral agency cannot be separated from the context in which actions occur. The concept of *relational agency* inextricably links moral action to the last assumption in Jameton's definition, constraints to action.

Critiques about the role of constraints arose early in the history of the concept. The first research on the experience of moral distress for nurses and the impact on patient care was conducted by Wilkinson [6]. Her research identified a gap in understanding about the location of constraints. Originally Jameton [3] identified constraints as institutional and external to the nurse. Wilkinson's model of moral distress acknowledged that contextual constraints might be real or *perceived*. Recognizing that constraints to action are sometimes perceived suggests that institutional constraints on action don't actually exist or that nurses who fail to take action are lacking in moral competency or knowledge, are powerless to take action, or may choose not to take action based on moral aptitude or character [40]. Our response to this critique is to point to the importance of nursing's, and other healthcare provider groups', increasing awareness that the experience of moral distress may occur as a process that evolves over time for many people [21, 22]. The consequence is that awareness of constraints and our ability to articulate what contributes to the experience occurs through reflection on professional values and obligations and therefore may evolve over time [21, 37].

Recently, nurse scholars have examined moral distress in novel ways in order to bring more theoretical depth to the concept. For example, Peter and Liaschenko [47] draw on feminist moral theory to provide an explanation of what might be happening in the experience of moral distress, and Lützén and Ewalds-Kvist [48] draw on Victor Frankl's work on meaning in an effort to bring theoretical depth to their own work on moral distress. In applying different philosophical lenses to the experience of moral distress, these authors are able to examine the assumptions present in Jameton's definition and move beyond a linear concept of moral distress to explore the complexity of enacting moral agency. For example, Peter and Liaschenko [47] suggest that moral agency is a socially connected phenomenon that encompasses identity, relationships, and responsibility, thereby surfacing aspects of constraints to moral agency that may be present, yet ambiguous and difficult to articulate.

As well, researchers acknowledge that constraints could be internal or external to the individual healthcare provider [49]. Newer definitions offered by researchers either do not explicitly identify the location of the constraints on action (e.g., see [17, 30, 32]) or are beginning to point to constraints as being located at the complex relational interplay between structures and agents [12, 17, 24]. Many of us studying moral distress have discussed moral agency and constraints as if they are separate ideas underpinned by different assumptions. While this is partially true, in this chapter we want to move forward by acknowledging that these two components of the definition (agency and constraints) are, in reality, inseparable. As such, it is imperative to understand the relationship between enacting moral agency and the elements that constrain moral agency

2.4   Appreciating the Reciprocity of Structure and Agency

Scholars and researchers in moral distress are increasingly calling for a *relational* approach to exploration of the concept of moral agency in order to better understand the complex relationships that exist between organizational structures and healthcare providers as moral agents. The assumptions we have pointed to above reflect implicit understandings about the *agency* of healthcare providers, as well as the *structures* they operate in and attempt to influence. In a traditional philosophical view of moral agency, we see " ...a person who is capable of deliberate action and/or who is in the process of deliberate action" [50, 51]. Further, "traditional perspectives on moral agency reflect a notion of individuals engaging in self-determining or self-expressive choice" [52] (see also [51]). Yet "moral agents in healthcare (patients, families, and professionals/providers) are not as 'equal' and autonomous as the traditional perspectives might assume" [51] (see also [43, 53]). This traditional view of moral agency has shifted over the past two decades as scholars have critiqued this view of agency as failing to acknowledge that agency is "enacted through *relationship* in particular *contexts*" [51]. In the context of healthcare, moral agency incorporates knowledge of such things as policies, protocols, unit and organizational culture and values, and interpersonal, human and material resources. Additionally, broad societal elements such as social, political, cultural, and economic values directly shape and influence both the healthcare environment and individual healthcare providers. Recognizing agency as relational moves decision-making about what actions are available to practitioners from the realm of the individual into the context in which the individual is operating and exposes the complexity that actually exists when someone chooses to act as a moral agent.

In moving decisions about moral agency from an individualistic perspective into a relational perspective, we want to move past the view of constraints resting either within the individual or with the organization. Rather, we believe moral agency and constraints reside at the *intersection* of structure and agent. We believe that structures, for example, sociopolitical and economic policies, influence decision-making at the micro, macro, and meso levels of healthcare delivery. The reverse is also true; individuals have the ability to influence sociopolitical and economic policies at these same levels. We are pointing to the idea of reciprocity between structure and agency, whereby individuals and organizations are in constant relationship with each other and therefore have the capacity to influence and be influenced by each other [19, 24, 54]. Sewell [55], a sociologist, describes the relationship between structure and agency as:

> Structures...are constituted by mutually sustaining cultural schemas and sets of resources that empower and constrain social action and tend to be reproduced by that action." Agents are empowered by structures, both by the knowledge of cultural schemas that enable them to mobilize resources and by the access to resources that enables them to enact schemas [55]

In using the word "empowered" to describe agents, Sewell's description of the relationship between structures and agents appears to overlook the fact that structures also have the capacity to disempower agents by constraining agency through restricting access to resources. Examples of restricting access to resources are evident in healthcare, such as when healthcare providers are excluded for discussions on resource allocation. However, there is also an assumption that all agents have some, albeit perhaps limited, access to resources and therefore have some capacity for agency.

Sewell's [55] understanding of the reciprocity that occurs at the intersection of structures and agents emphasizes the dynamic and evolving nature of structures, meaning that even small actions of moral agency have the potential to create change in the healthcare system. For example, nurses can work through their professional associations to advocate for more equitable allocation of healthcare resources. In initially naming and later refining the definition of moral distress, Jameton held moral agency as central to ameliorating or mitigating the experience [3, 26, 56]. Having said this, Jameton and others [54] recognize that action in the healthcare system is "essentially collaborative and collective" [26] requiring HCPs at all levels of the healthcare system to take action when they are confronted by ethical challenges that contribute to moral distress. Building from Jameton, we propose that moral distress be defined in relation to influences beyond those that would be considered institutional to broader sociopolitical contexts and not depend on the level of impossibility of action. By this, we mean that the definition of moral distress must be moved beyond the level of the individual. Toward this end, we point to the strength of the definition proposed by Varcoe and Pauly [17]:

> the experience of being seriously compromised as a moral agent in practicing in accordance with accepted professional values and standards. It is a relational experience shaped by multiple contexts, including the socio-political and cultural context of the workplace environment. (p. 59)

Conclusion

The work inspired by American philosopher Andrew Jameton's groundbreaking book on nursing ethics [3] continues to evolve. While more conceptual work is needed [7, 24], we certainly know enough to continue to improve the practice environments for nurses and other healthcare providers.

As we claimed earlier in this chapter, supporting healthcare providers as moral agents operating in complex organizational structures is prerequisite to offering effective and ethical healthcare and fostering a sustainable healthcare workforce. Our explorations in this paper have affirmed that the prevalence of moral distress is of significant concern. The expanding global interest in the topic means that we can continue moving the concept forward in order to help us have a more nuanced understanding of moral distress. A more nuanced understanding is foundational to supporting the well-being of healthcare providers so that they are in a position to more effectively deliver clinically and ethically sound healthcare.

This requires that we take action throughout our healthcare system, using a relational ethical perspective that attends to power dynamics across all levels [33], and the reciprocity that exists between structures and agents. At the individual level, healthcare providers ought to learn about how to deal with moral distress and how to develop moral resilience [57] early in their professional educational programs.[6] Further, healthcare providers would benefit from having supportive practice mentors assigned to encourage them as they initiate their practice. At the organizational level, leaders for healthcare practice ought to provide guidance that is visionary, innovative, and inspiring [58, 59]. Such guidance can encourage a values-based orientation to organizing practice environments so that the resources required to deliver clinically and morally sound care are more readily available.

For this values-based orientation to flourish, leaders and policy makers at larger systems levels should be inspired by a commitment to values rather than just the "bottom line" [33]. Indeed, it is our conviction that healthcare agencies, healthcare funders, and healthcare professional groups should operate according to a principle of "justice as shared responsibility" [60], where all those involved in healthcare delivery see improved healthcare, as well as reduced healthcare providers' moral distress as their shared moral goals. The widespread enactment of justice as shared responsibility would mean that resources were in place to promote the well-being of all involved in healthcare delivery—whether they are patients, families, communities, or healthcare providers.

References

1. Beauchamp T, Childress J. Principles of biomedical ethics. 6th ed. Oxford: Oxford University Press; 2008.
2. Rodney P, Burgess M, Phillips JC, McPherson G, Brown H. Our theoretical landscape. In: Storch JL, Rodney P, Starzomski R, editors. Toward a moral horizon. 2nd ed. Don Mills: Pearson; 2013.
3. Jameton A. Nursing practice: the ethical issues. Prentice-Hall: Englewood Cliffs; 1984.
4. Fry ST, Harvey RM, Hurley AC, Foley BJ. Development of a model of moral distress in military nursing. Nurs Ethics. 2002;9(4):373.
5. Hamric AB. Moral distress in everyday ethics. Nurs Outlook. 2000;48(5):199–201.
6. Wilkinson JM. Moral distress in nursing practice: experience and effect. Nurs Forum. 1987;23(1):16–29.
7. Rodney PA. What we know about moral distress. Am J Nurs. 2017;117(2 Suppl 1):S7–S10.
8. Taylor C. Nationalism and modernity. In: The morality of nationalism. New York: Oxford University Press; 1997. p. 31.
9. Rodney PA. Seeing ourselves as moral agents in relation to our organizational and sociopolitical contexts. J Bioeth Inq. 2013;10(2):313–5.
10. Burston AS, Tuckett AG. Moral distress in nursing: contributing factors, outcomes and interventions. Nurs Ethics. 2013;20(3):312–24.
11. Crane MF, Bayl-Smith P, Cartmill J. A recommendation for expanding the definition of moral distress experienced in the workplace. Aust N Z J Organ Psychol. 2013;6:e1.

[6] For an itemization of theoretically grounded and practical suggestions to develop moral resiliency in healthcare providers, we refer readers to the recent Rushton et al. [57] article, *A collaborative state of the science initiative: Transforming moral distress into moral resiliency in nursing.*

12. Ramos FRS, Barlem ELD, Brito MJM, Vargas MA, Schneider DG, Brehmer LCF. Conceptual framework for the study of moral distress in nurses. Texto Contexto Enferm. 2016;25(2)
13. Austin W. The ethics of everyday practice: healthcare environments as moral communities. Adv Nurs Sci. 2007;30(1):81.
14. McCarthy J, Deady R. Moral distress reconsidered. Nurs Ethics. 2008;15(2):254–62.
15. Shoorideh FA, Ashktorab T, Yaghmaei F, Alavi Majd H. Relationship between ICU nurses' moral distress with burnout and anticipated turnover. Nurs Ethics. 2015;22(1):64–76.
16. Baylis F, Kenny NP, Sherwin S. A relational account of public health ethics. Public Health Ethics. 2008. https://doi.org/10.1093/phe/phn025.
17. Varcoe C, Pauly B, Webster G, Storch J. Moral distress: tensions as springboards for action. HEC Forum. 2012;24(1):51–62.
18. Hoffmaster B. Introduction. In: Hoffmaster BF, editor. Bioethics in social context. Philadelphia: Temple University Press; 2001. p. 1–11.
19. Hanna DR. Moral distress: the state of the science. Res Theory Nurs Pract. 2004;18(1):73.
20. Musto L, Rodney P, Vanderheide R. Toward interventions in moral distress: navigating reciprocity between structure and agency. Nurs Ethics. 2015;22(1):91–102.
21. Campbell SM, Ulrich CM, Grady C. A broader understanding of moral distress. Am J Bioeth. 2016;16(12):2.
22. Webster G, Baylis FE. Moral residue. In: Rubin SB, Zoloth L, editors. Margin of error: the ethics of mistakes in the practice of medicine. Hagerstown: University Publishing Group; 2000. p. 13.
23. Austin W, Lemermeyer G, Goldberg L, Bergum V, Johnson MS. Moral distress in healthcare practice: the situation of nurses. HEC Forum. 2005;17(1):33–48.
24. Musto LC, Rodney PA. Moving from conceptual ambiguity to knowledgeable action: using a critical realist approach to studying moral distress: critical realism and moral distress. Nurs Philos. 2016;17(2):75–87.
25. Fourie C. Moral distress and moral conflict in clinical ethics. Bioethics. 2015;29(2):91–7.
26. Jameton A. Dilemmas of moral distress: moral responsibility and nursing practice. AWHONNS Clin Issues Perinat Womens Health Nurs. 1993;4(4):542.
27. Pauly BM, Varcoe C, Storch J. Framing the issues: moral distress in health care. HEC Forum. 2012;24(1):1–11.
28. Hanna DR. Moral distress redefined: the lived experience of moral distress of nurses who participated in legal, elective, surgically induced abortions. A dissertation; 2002.
29. Austin WA, Bergum V, Goldberg L. Unable to answer the call of our patients: mental health nurses' experience of moral distress. Nurs Inq. 2003;10(3):177–83.
30. Kälvemark S, Hoglund AT, Hansson MG, Westerholm P, Arnetz B. Living with conflicts-ethical dilemmas and moral distress in the health care system. Soc Sci Med. 2004. https://doi.org/10.1016/S0277-9536903000279-X.
31. Nathaniel AK. Moral reckoning in nursing. West J Nurs Res. 2006;28(4):419–38. https://doi.org/10.1177/0193945905284727.
32. Mitton C, Peacock S, Storch J, Smith N, Cornelissen E. Moral distress among health system managers: exploratory research in two British Columbian health authorities. Health Care Anal. 2011. https://doi.org/10.1007/s10728-010-0145-9.
33. Rodney P, Harrigan M, Jiwani B, Burgess M, Phillips JC. A further landscape: ethics in health care organizations and health/health care policy. In: Storch J, Rodney P, Starzomski R, editors. Toward a moral horizon. 2nd ed. Toronto: Pearson; 2013. p. 25.
34. Barlem E, Ramos F. Constructing a theoretical model of moral distress. Nurs Ethics. 2014;22(5):608–15.
35. Musto L. Risking vulnerability: enacting moral agency in the is/ought gap [Unpublished doctoral dissertation]; 2018.
36. Austin WJ, Kagan L, Rankel M, Bergum V. The balancing act: psychiatrists' experience of moral distress. Med Health Care Philos. 2008;11(1):89–97.
37. Carse A. Moral distress and moral disempowerment. Narrat Inq Bioeth. 2013;3(2):147.
38. Mack C. When moral uncertainty becomes moral distress. Narrat Inq Bioeth. 2013;3(2):106–9.

39. McCarthy J, Gastmans C. Moral distress: a review of the argument-based nursing ethics literature. Nurs Ethics. 2014. https://doi.org/10.1177/0969733014557139.
40. Johnstone MJ, Hutchinson A. 'Moral distress' – time to abandon a flawed nursing construct? Nurs Ethics. 2013;22(1):5–14.
41. Repenshek M. Moral distress: inability to act or discomfort with moral subjectivity. Nurs Ethics. 2009. https://doi.org/10.1177/0969733009342138.
42. Paley J. Commentary: the discourse of moral suffering. J Adv Nurs. 2004;47(4):364–5.
43. Varcoe C, Pauly B, Storch JL, Newton L, Makaroff KS. Nurses' perception of and responses to moral distress. Nurs Ethics. 2012;19(4):12.
44. Sundin-Huard D, Fahy K. Moral distress, advocacy and burnout: theorising the relationships. Int J Nurs Pract. 1999;5(1):8–13.
45. Musto L, Schreiber RS. Doing the best I can do: moral distress in adolescent mental health nursing. Issues Ment Health Nurs. 2012;33(3):137.
46. Rodney P, Varcoe C. Constrained agency: the social structure of nurses' work. In: Bayis F, Hoffmaster B, Sherwin S, Borgerson K, editors. Health care ethics in Canada. 3rd ed. Toronto: Nelson; 2012. p. 17.
47. Peter E, Liaschenko J. Moral distress reexamined: a feminist interpretation of nurses' identities, relationships, and responsibilites. J Bioeth Inq. 2013;10(3):337–45.
48. Lützén K, Ewalds-Kvist B. Moral distress and its interconnection with moral sensitivity and moral resilience: viewed from the philosophy of viktor e. Frankl. J Bioeth Inq. 2013;10(3):317–24.
49. Austin W, Rankel M, Kagan L, Bergum V, Lemermeyer G. To stay or to go, to speak or stay silent, to act or not to act: moral distress as experienced by psychologists. Ethics Behav. 2005;15(3):197–212.
50. Angeles PA. Dictionary of philosophy. New York: Barnes & Noble; 1981.
51. Rodney P, Kadyschuk S, Liaschenko J, Brown H, Musto L, Snyder N. Moral agency: relational connections and supports. In: Storch J, Rodney P, Starzomski R, editors. Toward a moral horizon: nursing ethics for leadership and practice. 2nd ed. Don Mills: Pearson; 2013. p. 27.
52. Taylor C. Multiculturalism and "The politics of recognition"; with a commentary by Amy Gutmann. Princeton: Princeton University Press; 1992.
53. Peter E. Fostering social justice: the possibilities of a socially connected model of moral agency. Can J Nurs Res. 2011;43(2):11–7.
54. Hartrick-Doane G, Varcoe C. Relational practice and nursing obligations. In: Storch JL, Rodney P, Starzomski R, editors. Toward a moral horizon. 2nd ed. Don Mills: Pearson; 2013. p. 27.
55. Sewell WHA. Theory of structure: duality, agency, and transformation. Am J Sociol. 1992;98(1):1–29.
56. Jameton A. A reflection on moral distress in nursing together with a current application of the concept. Bioeth Inq. 2013;10:297–308.
57. Rushton CH, Schoonover-Shoffner K, Kennedy MS. Executive summary: transforming moral distress into moral resilience in nursing. Am J Nurs. 2017;117(2):52–6.
58. Curtis EA, de Vries J, Sheerin FK. Developing leadership in nursing: exploring core factors. Br J Nurs. 2011;20(5):306–9.
59. Gaudine A, Lamb M. Nursing leadership and management. Upper Saddle River: Pearson Education; 2014.
60. Young IM. Responsibility for justice. New York: Oxford University Press; 2010.

Anne J. Davis, Marsha Fowler, Sophia Fantus,
Joseph J. Fins, Michelle Joy, Katherine Kruse,
Alyssa Burgart, Margaret Lindsey, Kim Mooney-Doyle,
Tanya Uritsky, and Christine Grady

A.J. Davis
University of California, San Francisco, American Academy of Nursing "Living Legend" in
Nursing, San Francisco, CA, USA

M. Fowler
Department of Ethics, Azusa Pacific University, Azusa, CA, USA

S. Fantus
Center for Medical Ethics and Health Policy, Baylor College of Medicine, Houston, TX, USA

J.J. Fins
Division of Medical Ethics, Weill Cornell Medical College, New York, NY, USA

New York Presbyterian Hospital Weill Cornell Medical Center, New York, NY, USA

Consortium for the Advanced Study of Brain Injury, Weill Cornell and Rockefeller
University, New York, NY, USA

Solomon Center for Health, Law and Policy, Yale Law School, New Haven, CT, USA

M. Joy
Hospital of the University of Pennsylvania, Philadelphia, PA, USA

K. Kruse
Pediatric Critical Care, Children's Respiratory and Critical Care Specialists, Minneapolis,
MN, USA

Department of Clinical Ethics, D, Children's Hospital and Clinics of Minnesota, Minneapolis,
MN, USA

A. Burgart
Department of Anesthesiology, Perioperative, and Pain Management,
Lucile Packard Children's Hospital at Stanford, Stanford Center for Biomedical Ethics,
Stanford University, Stanford, CA, USA

M. Lindsey
Board Certified Chaplain, Association of Professional Chaplains, State College, PA, USA

© Springer International Publishing AG 2018
C.M. Ulrich, C. Grady (eds.), *Moral Distress in the Health Professions*,
https://doi.org/10.1007/978-3-319-64626-8_3

K. Mooney-Doyle
University of Maryland School of Nursing, Baltimore, MD, USA

T. Uritsky
Pain Management and Palliative Care, Hospital of the University of Pennsylvania,
Philadelphia, PA, USA

C. Grady (✉)
Department of Bioethics, National Institutes of Health, Bethesda, MD, USA
e-mail: CGrady@cc.nih.gov

Most healthcare professionals can describe one or more clinical situations which left them feeling at best unsure and at worst troubled or distressed by the outcome or the decisions that were made for a patient. In some of these situations, healthcare professionals experienced moral distress because of their involvement in what they perceived as a morally undesirable situation. Moral distress is a phenomenon experienced by a wide range of healthcare professionals in a variety of settings. As described elsewhere in this book, although the term was first used to describe an experience of nurses, who often felt constrained by rules or hierarchies in healthcare facilities, healthcare professionals across the spectrum have experienced and described moral distress. Often, the themes and experiences of moral distress are similar and shared across healthcare professions, themes of lack of control, powerlessness, and unrealistic expectations. Yet, differences in context and role responsibilities influence and can result in dissimilar experiences of moral distress in healthcare providers with different roles and responsibilities. For some healthcare professions, the phenomenon of moral distress has not been named or recognized or has been called something else.

In this chapter, we have brought together the voices of multiple healthcare professionals to describe moral distress from their own perspectives. Nurses, physicians, a social worker, a chaplain, and a pharmacist, each provide thoughtful insights into how they understand and have experienced moral distress from their own disciplinary perspective. Two of the contributions found in Part 1 offer an academic survey of the literature, including the history and current thinking and research on moral distress in nursing (Davis and Fowler) and in social work (Fantus). The other contributions in this chapter collection found in Part 2 (in alphabetical order by Fins, Joy, Kruse and Burgart, Lindsey, Mooney, Uritsky) offer a more personal perspective on how they, as physician, psychiatrist, physicians, chaplain, pediatric nurse, and pharmacist, respectively, experience or witness moral distress in their practice. Many describe the confluence of disparate factors that result in or confound the experience of moral distress, and many describe the value of story-telling, sharing experiences, and thinking or acting together to convert distress-inducing situations into possible opportunities for growth and change.

Part 1 Contributors

Anne Davis and Marsha Fowler, *Moral Distress in Nursing: Looking Back to Move Forward*

Sophia Fantus, *Social Work Perspective: Moral Distress*

Part 2 Contributors

Joseph Fins, *A Source of Moral Distress: The Corporatization of Medicine*

Michelle Joy, *Moral Distress: A Psychiatrist perspective*

Katherine Kruse and Alyssa Burgart, *Physicians' Experiences of Moral Distress and Burnout*

Reverend Peggy Lindsey, *A Chaplain's Perspective on Moral Distress*

Kim Mooney-Doyle, *Moral distress in pediatric nursing and research*

Tanya J. Uritsky, *Pharmacist's Perspective on Moral Distress in Palliative Care: A narrative*

3.1   Reviews of Moral Distress in Nursing and in Social Work

3.1.1   Moral Distress in Nursing: Looking Back to Move Forward

Anne J. Davis and Marsha D. Fowler

3.1.1.1 Introduction

No moral issue is historically context-free, and that includes the issue of moral distress. Most discussion of moral distress reaches back to Jameton's [1] work *Nursing Practice: The Ethical Issues*. Here, Jameton identifies three categories of moral concern: moral uncertainty, moral dilemma, and moral distress. *Moral uncertainty* occurs when the nurse is unclear whether a difficult situation is moral in itself, and if it is, what values or obligations are being challenged. Moral uncertainty is amenable to ethics education and to professional socialization and education into the values of nursing. *Moral dilemma* occurs when the nurse faces a conflict of values or of ethical obligations, which presents alternative and conflicting courses of action. Ethics education is important here as well, as is participation in the community of moral discourse. *Moral distress*, in its original definition, "arises when one knows the right thing to do, but institutional constraints make it nearly impossible to pursue the right course of action." [1]. Moral distress is not amenable to ethics education; such education simply helps the nurse be clearer and more articulate about the nature of the distress, which is not a bad thing in itself. In moral distress the nurse has a degree of certainty about the right action to take and yet is obstructed from taking that action by the institutional constraints, later given some nuance and modified by Jameton to distinguish between internal and external constraints [2]. The *distress* of *moral distress* is not generated by discomfort intrinsic to the case circumstances. It arises from external, specifically institutional, barricades to morally right action, or to internal social and psychological factors such as fear of job loss, self-doubt, timidity, trepidation, socialization against questioning medical orders, etc. What is important beneath Jameton's articulation of these three responses to moral issues is that his perspective was written from a nurses' point of view. The opening words of his work are

> Nursing is the morally central health profession. Philosophies of nursing, not medicine, should determine the image of healthcare and its future directions. In its anxiety to control the institutions and technology of healthcare, medicine has allowed the central values of healthcare—health and compassion—to fall to the hands of nurses. Nurses thus supply the real inspiration and hope for progress in healthcare, and among health professionals, represent the least equivocal commitment to their clientele [1, p. xvi].

He is correct of course. While we are grateful for Andy's friendship and colleagueship over the decades, it is important to acknowledge how great a debt of gratitude nursing owes to him. Among the philosophers who were early into the rise of bioethics movement, Dr. Jameton was distinctive. While other philosophers were uniformly medically identified, Jameton encamped with nursing. He provided second wave nursing ethics (post-1965) with rigorous, groundbreaking, and creative scholarship.

Continuing research has given rise to a host of corollary terms including moral outrage, moral courage, moral resilience, moral residue, moral sensitivity, and more. There has been an escalation of interest, since 2010, in moral distress in the nursing research literature, across settings, roles, and even countries. Nursing databases such as CINAHL include research articles on moral distress from Canada, Ireland, Italy, the Netherlands, Brazil, Malawi, Australia, Greece, Turkey, Belgium, New Zealand, Switzerland, Iran, Jordan, Sweden, Israel, Uganda, and more. In short, moral distress in nursing circles the globe.

3.1.1.2 The Moral Milieu of an Institution: Predecessor Literature

In discussions of moral distress, greater attention is often accorded to the moral environment in which nursing is practiced. Canadian nurse-ethicist Storch and colleagues note:

> Within Canada's fast-paced, ever-changing healthcare environment, providers are experiencing difficulty practising according to their professional ethical standards, leading many to experience moral or ethical distress. Limited attention has been paid to improvements in the ethical climate in healthcare settings in research focusing on nurses' workplaces. [3]

Storch points to the research neglect of workplace ethical climate. While she is correct regarding contemporary neglect of research on the moral environment of healthcare, there is a precursor literature that, for the most part, remains neglected as well. This literature has much to contribute to the elucidation of both the historical development of the concept of moral distress, its clinical expression, and its broader clinical, institutional, and social context that predates Jameton's definition.

For example, Davis and Aroskar's book *Ethical Dilemmas and Nursing Practice* [4], now in its fifth edition, devotes an entire chapter in each successive edition to social, institutional, and professional factors that make "being moral" difficult [4]. Several distinctive features in the first edition set it apart from prior and succeeding literature. While Davis and Aroskar were not the first to point to institutional structures that affect the practice of nursing, they were to first to explicitly set it within an ethical nexus for analysis. In answering their question of whether or not nurses could be moral in their practice, they looked to broad social issues such as the social location of women and gendered social roles reinforced by institutional structures and embedded in law, as well as to the suffusion of American healthcare by a business model; to institutional issues, specifically to the nurse's competing loyalties to hospital, physician, and patient, but also to hierarchical structures of authority and communication; and to the historically rooted nurse—physician relationship with its sex-stereotyped roles, authority, communication, and medical paternalism. Here, Davis and Aroskar cite Leonard Stein's highly original work "The Doctor—Nurse Game," which had great explanatory power in its day, and still does to some extent [5–7]. Another distinctive feature is that Davis and Aroskar moved significantly beyond the "dilemma-based" and "principle-based" expositions of that day to look at the broader context of ethics in nursing. They maintain that

> The overriding ethical issue for nurses, especially those working in hospitals, can best be described as one of multiple ethical obligations coupled with the question of authority… The physician has a special legal relationship with the patient, whereas the nurse's legal obligations vary according to a state's nurse-practice act. This fact makes nursing ethics more complex in clinical settings….Such issues as professional role, gender, education, public image, work environment, and status are central to nursing history and its present situation in which ethical dilemmas occur. [8]

Davis and Aroskar were not the first to deal with institutional constraints to nursing practice. For example, Marlene Kramer's [9] work *Reality Shock: Why Nurses Leave Nursing* presents her sociological research on the entry of new nursing graduates into nursing practice. She focuses on "the discrepancy and the shock like reactions that follow when the aspirant professional perceives that many professional ideals and values are not operational and go unrewarded in the work setting." [2]. She looks at "the seeds of discontent," that is, the effects of values discordance on new nurses and the value structure of the nursing profession against the value structure of bureaucratic healthcare that employs the nurse. She writes "the goal of adaptation in a reality shock situation is the creation of a viable habitat in which one can be productive, effective, and content for a longer and probably indefinite period of time." [9]. While Kramer's method is decidedly sociological, the cases that she cites demonstrate challenges to the nurses'—or nursing's—values and ideals; the cases are intrinsically ethical but not named as such. It is left to Davis and Aroskar to take these seeds of discontent and to exegete them as explicitly ethical in nature, and to set them within a larger social context beyond that of the particular healthcare institution.

Vaughn's research for her 1935 master's degree "dissertation" (thesis) *The Actual Incidence of Moral Problems in Nursing: A Preliminary Study in Empirical Ethics* is the first, pre-bioethics, ethics research that addresses moral issues in practice [10]. It is the first research that touches upon institutional strictures and conditions that trouble or fetter nurses. The object of her research was "to obtain, by means of diaries, the actual incidence of moral problems occurring among nurses." [10]. A total of 95 nurses returned 288 diaries, kept over a period of three months, and that "yielded 2,265 moral problems" categorized using a modification of Lehmkuhl's classification of moral problems [10, 11]. Vaughn reports that she divided the data "into three general classes: moral problems, cultural problems, and questions that do not seem to imply problems. The morally-involved problems...exceed out of all proportion the balance of the material..." [10]. The vast majority of incidents or concerns recorded were in fact of a moral nature; the nurse–physician relationship was chief among all of the incidents reported. "The greater number of these involved questions of the propriety of the nurses making suggestions to the doctor regarding his orders, questions of loyalty to the physician, and doubts in matters of making hospital rounds with the doctor." [10]. More specifically these narratives included questions of the nurse's moral responsibility where physicians ordered wrong treatments, physicians made life-threatening mistakes, a physician declined to use gloves in multiple pelvic exams, a nurse recognizing a broken bone that the physician does not, and so on. In the recorded cases, nurses were aware of what was or was not an ethical issue and where the crux of the issue lay. That is, their uncertainty was not over the moral nature or content of the concern, or over what would be an appropriate moral outcome, but over how to proceed.

Vaughn uses Lehmkuhl's classification system:

> ...with some slight modifications. Lehmkuhl's "duties to state" was omitted and, for our purpose, "duties to the patient", "duties to the hospital", and "duties to the profession" supplied [11, page 19]

She thus incorporates but does not analyze some of the competing ethical obligations that Davis and Aroskar articulate some years later. Vaughn does, of course, predate the critical theories to which nursing ethics scholars have access today, specifically those that examine intersections of race, gender, and power.

Though hers is the first actual research, certainly nurses' distress over institutional practices, physician-related issues, and more predates Vaughn's work. In 1928, Sara Parsons writes in her ethics textbook, *Nursing Problems and Obligations*,

> A nurse may find herself in an institution where she cannot respect her superior officers or approve of the policy of the institution; if so, she can hardly stay in such a place permanently without conforming to the objectionable ways or seeming to condone the malpractice of other officials....She must in all situations try to keep a clean-cut ideal of honor for herself [12, p. 109]

This position is mirrored in the 2015 ANA Code, though some have mistakenly believed that this hardline is new to nursing ethics. It is over 100 years old and enduring. The passage reads:

> Nurses should address concerns about the healthcare environment through appropriate channels and/or regulatory or accrediting bodies. After repeated efforts to bring about change, nurses have a duty to resign from healthcare facilities, agencies, or institutions where there are sustained patterns of violation of patient's rights, where nurses are required to compromise standards of practice or personal integrity, or where the administration is unresponsive to nurses' expressions of concern. Following resignation, reasonable efforts

to address violations should continue. … By remaining in such an environment, even if from financial necessity, nurses risk becoming complicit in ethically unacceptable practices and may suffer adverse personal and professional consequences [13, pp. 24–25].

Institutional constraints, including difficult relationships with physicians are noted in the nursing ethical literature from the 1870s onward. Isabel Robb, in *Nursing Ethics: For Hospital and Private Use* [14], devotes a section to the "relation of the nurse to the physician." In that section she acknowledges that some physicians are "incapable," and that

> …if truth be told, there are rare instances in which the physician is unworthy of the respect both of nurse and patient, and the former, when she has gone through one such unsatisfactory experience, is fully justified in avoiding the care of patients under his charge….but although the nurse may be longsuffering…she is not expected to put up with unjust or rude behavior, and when she finds that, through no fault of hers and despite her best endeavors, she cannot work in harmony with the physician, she is fully justified in leaving the case as soon as an efficient substitute has been found to take her place. [14]

In Hirschman's typology of exit (leave), loyalty (stay), and voice (express concern) the historical weight of nursing's ethical literature, both heritage and contemporary ethics literature, is to attempt to make institutional change to improve the moral milieu or its policies, but where the institution is refractory to change—exit, or exit with voice [15, p. 54].

3.1.1.3 Shaping and Re-Shaping the Moral Milieu

As Davis and Aroskar note, effective moral navigation requires formal mechanisms for discussing ethical dilemmas as well as a socio-ethical culture that will foster discussion and action [8]. Intrinsic to this, of course, is a need for foundational ethics education. However, ethics education cuts both ways in that it assists in analysis and decision-making and can bring a clarity—that can make the issues even more acute and frustrating. The 2015 American Nurses *Association Code of Ethics for Nurses with Interpretive Statements* (Code) addresses the moral environment of nursing. It states:

> Nurses are responsible for contributing to a moral environment that demands respectful interactions among colleagues, mutual peer support, and open identification of difficult issues, which includes ongoing professional development of staff in ethical problem solving. Nurse executives have a particular responsibility to assure that employees are treated fairly and justly, and that nurses are involved in decisions related to their practice and working conditions. Unsafe or inappropriate activities or practices must not be condoned or allowed to persist. Organizational changes are difficult to achieve and require persistent, often collective efforts over time. Participation in collective and inter-professional efforts for workplace advocacy to address conditions of employment is appropriate [13].

Beyond the immediate practice context, the Code looks to shaping the moral environment through regulatory and accrediting bodies, and professional associations. It states:

> The workplace must be a morally good environment to ensure ongoing safe, quality patient care and professional satisfaction for nurses and to minimize and address moral distress,

strain, and dissonance. Through professional organizations, nurses can help to secure the just economic and general welfare of nurses, safe practice environments, and a balance of interests. These organizations advocate for nurses by supporting legislation; publishing position statements; maintaining standards of practice; and monitoring social, professional, and healthcare changes [13].

The Code also assists nursing in navigating obligations, at least in terms of prioritizing the nurse's commitment to the patient above other obligations. Provision two states: "The nurse's primary commitment is to the patient, whether an individual, family, group, community or population" [13]. The general conflict of obligations or values envisioned are those between nurse and physician, nurse and employer, nurse and nurse. At times, however, it will include a conflict of obligations or values between nurse and patient. The Code does provide a range of mechanisms for nurses to safeguard their own moral integrity without compromising patient care, such as withdrawing from situations where there is a conflict of interest, to engage in conscientious objection, to identify prior to employment limits to practice, and more.

3.1.1.4 Ethics Education and Enduring Issues in Nursing

There are enduring, structural, issues within nursing itself that are antecedents of moral distress. For example, multiple entry points in nursing disadvantage both the profession and individual nurses. As Donley notes,

Registered nurses are undereducated members of the healthcare team, when compared with physicians, social workers, physical therapists, pharmacists, and dieticians to name a few. Looking beyond the clinical environment, the nurse work force also lacks the educational credentials of persons in the business, investor, and insurance communities that now play significant roles in healthcare decisions. Under-educated members of the health team rarely sit at policy tables or are invited to participate as members of governing boards. Consequently, there is little opportunity for the majority of practicing nurses to engage in clinical or healthcare policy [16].

As has been called for decades, standardizing the entry level for nursing practice at the baccalaureate level and above would go a long way toward securing the nurse's place at the table, if one means of addressing moral distress requires nurse participation in clinical policy making.

Early modern nursing, 1870s to 1965, (that is, prior to the American Nurses Association 1965 position paper on nursing education [17]), viewed ethics education, the moral formation of the student, and the "tone" of the school and hospital (i.e., moral milieu), as important as clinical and scientific content. From 1900 to 1965, at any given point in time, there were from two to 11 textbooks on nursing ethics widely available to schools of nursing. These books were largely lost when nursing moved from hospital-based education into colleges and universities, and ethics education shifted from nursing schools into departments of philosophy or theology (or is lost altogether). This shift was concurrent with the rise in the field of bioethics, which nursing embraced, substituting it for the 125 years of nursing ethics that had developed [18, 19].

A second issue, then, is that participation in moral policy making requires ethics education, a persistent problem in contemporary nursing education. As central as ethics is to nursing practice, it is often treated as peripheral to the curriculum, to be squeezed in if there is space. It is also treated as if it were not an academic discipline in itself, but rather the domain of commonsense and conscience (or the domain of privatized personal opinion) that does not require formal knowledge or competence to qualify to teach it. Were the centrality of nursing ethics to the profession, to professional formation, and to professional practice to be reclaimed [19], nurses would be better equipped to face moral concerns in professional practice and to navigate them appropriately.

A third example of a structural issue within nursing relates to larger trends in higher education that affect nursing (and medical) education: the failure to educate nurses for "civic professionalism" in general but more strongly so with the corporatization of education and the mantra of "job ready, career ready" with a devaluing of a broader education that includes the liberal arts, the humanities [20]. The humanities, as a reflection upon human experience, equip learners to assess and challenge the social and political—and institutional—status quo. Sullivan, in *Work and Integrity: The Crisis and Promise of Professionalism in America*, claims that "the narrowing of professional claims toward the purely cognitive or technical in recent decades has contributed to the weakening of professionalism," [21] resulting in a decline in professional civic engagement, a loss of concern for the welfare of society, a decline in altruism and professional ethics, and the reconceptualization of the recipient as consumer of a commodity. In his work with Benner, they call for several changes to nursing education to move professionals away from technical and into civic professionalism, a move that would ultimately strengthen professional ethics for practice and ethical comportment [20, 22].

While we have principally addressed the antecedent and extant literature that looked toward external constraints to moral action, internal constraints to moral action exist as a source of moral distress. Sullivan, drawing heavily upon the work of Benner, supplies one corrective approach to those internal constraints by suggesting ways in which to reshape nursing education. He writes that

> …the essential goal of the professional school must be to form practitioners who are aware of what it takes to become competent in their chosen domain and equip them with the reflective capacity and motivation to pursue genuine expertise. In the case of nursing, for example, this would mean studying and understanding the changing conditions of practice, as illuminated by history and the social sciences, alongside the study of the field's particular knowledge base in the physical sciences. …Identification and formation of skillful ethical comportment must be the organizer of competence and inspiration of expert work [21, 22]

Professional formation, developed competence, reflective capacity, motivation for expertise, sociopolitical understanding; knowledge of history, physical, and social sciences; disciplinary knowledge and expertise…are all to the end of "identification and formation of skillful ethical comportment as the organizer of competence and inspiration of expert work."

3.1.1.5 Various Critiques of Moral Distress

As a concept, moral distress, for all its currency and favor, is not itself without controversy. In their provocative article "'Moral Distress' – time to abandon a flawed nursing construct?" Johnstone and Hutchinson posit major—fatal—flaws with the concept of *moral distress*. They write:

> Essential to the theory of moral distress is the assumption that such a state of distress in fact exists. Much of what has been written about moral distress, however, involves little more than an appropriation of 'ordinary' psychological and emotional reactions (e.g. frustration, anger, anxiety, dissatisfaction) that nurses may justifiably feel when encountering difficult ethical issues, disagreements and conflicts in the workplace. Whether these reactions necessarily constitute 'moral' distress, however, is debatable [23, p. 10]

Vaughn's diary cases would seem to support Johnstone and Hutchinsons's contention regarding specifically *moral* distress. The 2265 cases represent a variety of psychological responses to moral situations, but not *moral distress* per se. Indeed, the nurses seem to be particularly morally hale and hardy.

Johnstone and Hutchinson also address the issue of nurses' moral decision-making competence. They write:

> Linchpin to the theory of moral distress is the idea that nurses know the right thing to do but are unable to carry it out. This idea is highly questionable on at least three accounts: first, it assumes, without supporting evidence, the unequivocal correctness and justification of nurses' moral judgments in given situations (rarely are the bases of the nurses' moral judgments revealed, and rarely is it admitted that nurses might be mistaken or misguided in their moral judgments, or that their moral judgments may be just plain wrong) ...[23]

Vaughn's cases contravene Johnstone and Hutchinson in that the 2265 cases do not generally present situations where right and wrong are unclear, uncertain, or truly in question (e.g., a physician instructs a nurse to deceive a patient, a nurse is ordered to falsify a patient record, the nurse is sexually harassed by the physician, and the like). These are cases of *moral failure*, not moral uncertainty. In the overwhelming majority of Vaughn's cases the *right* and the *good* are crystal clear; what is uncertain is *moral navigation* in the situation.

While it is possible to separate psychological and moral distress in practice, little conceptual work has been done on the actual distinction. For example, Noelle, at 24, was terminal from cystic fibrosis and after one-too-many end-of-life struggles, requested to be extubated. Extubation and her death within minutes was cause for psychological distress and grief for the loss of a long-time patient, but not for moral distress. Granted, the values of respect for the patient's wishes, and the odiousness of prolonging or even enhancing suffering in the face of ineluctable terminality were hardened values in this case. Previous to this instance, however, when the physician ignored Noelle's enduring, stated position, and intubated her anyway, the nurse was beset by both psychological and moral distress. While it may not be possible to do so, the failure to distinguish between psychological and moral distress runs the risk of conflating the two so that all psychological distress is labeled moral distress.

As noted above, ethics education can simultaneously clarify and aggravate moral distress. In Fowler and Mahon's 1979 research, students educated in clinical bioethics were able accurately to identify and parse actual moral issues and dilemmas in their clinical practice. But, having acquired the necessary ethical knowledge and analytical skill, they consequently articulated a level of *moral outrage* (as I had termed it). This was not distress, in the sense of moral suffering, moral impotence, oppression, or even victimization. Rather it was moral outrage borne of knowledge and ethical decision-making skill—strength and empowerment—not weakness and powerlessness [24].

McCarthy and Gastmans, [25] McCarthy and Deady, [26] Johnstone and Hutchinson [23] Peter and Liaschenko et al. [27] and others have provided trenchant critiques of the concept of moral distress as it has developed since 1984. Given that contemporary nurses find the concept of moral distress to have explanatory power, and given that the concept itself is problematic, Peter makes this observation:

> It may be that moral distress has made the social–moral space of nursing expressible in a way that many other concepts have not, with moral distress acting as window through which nurses can identify and describe the ethical nuances of their experiences. The problem may be, however, that we have asked too much of this concept by attempting to articulate more about the nature of nurses' ethical lives than it can reliably hold which has led to confusion regarding the meaning of moral distress and an over-emphasis on nurses' weaknesses as opposed to their strengths. My first recommendation, therefore, is that we also highlight alternative concepts in nursing ethics or develop, adapt, or borrow new ones that speak to the social–moral space of nurses. It is not that moral distress is no longer relevant, but we need to expand our understanding through additional concepts that help us understand the ethics of nursing work with its frequent proximity to patients or clients and its political positioning in a variety of settings. After all, the social–moral space of nurses does not just generate distress; it also opens opportunities to improve the well-being of patients because nurses are often in the position to provide and coordinate care in a way that recognizes patients as unique people [28].

It seems, then, that three things are needed. First, the very concept of moral distress needs to be subject to greater conceptual rigor and development. Second, nursing must address internal constraints of moral action, particularly those aspects of nursing professional formation and education that shape nurses' apparatus for ethical analysis and decision-making and equip and strengthen them for moral navigation in the contemporary healthcare system. Third, nursing as a profession, through its professional associations must continue to engage in social criticism and sociopolitical activism for social change, not simply on behalf of those who need advocacy, but for the larger social good that encompasses human health, rights, dignity, well-being, and flourishing. That social good includes the natural world in which humanity is situated; תיקון עולם (*tikkun olam*), for the repair of the world, for the healing of the world. Here we would concur with Jameton's extension of the concept of moral distress to encompass inter-connectedness, respect for all life, equality, and modesty of consumption [29]. Greater attention needs to be paid to the consanguinity of humanity; its place in and responsibility for the larger social, political, and physical environment; and the rights of the non-human world over against humanity. As the current (2015) *Code of Ethics for Nurses with Interpretive Statements* notes,

> Social justice extends beyond human health and well-being to the health and well-being of the natural world. Human life and health are profoundly affected by the state of the natural world that surrounds us. Consistent with Florence Nightingale's historic concerns for

environmental influences on health, and with the metaparadigm of nursing, the profession's advocacy for social justice extends to eco-justice. Environmental degradation, aridification, earth resources exploitation, ecosystem destruction, waste, and other environmental assaults disproportionately affect the health of the poor and ultimately affect the health of all humanity. Nursing must also advocate for policies, programs, and practices within the healthcare environment that maintain, sustain, and repair the natural world. As nursing seeks to promote and restore health, prevent illness and injury, and alleviate pain and suffering, it does so within the holistic context of healing the world [13, p. 37].

We call for greater rigor and clarity for the concept of moral distress, but also for its extension beyond our own anthropocentric distress.

3.1.1.6 Conclusion

In the end, enduring issues affecting nursing cannot finally be extracted from the larger, encompassing social-structural issues that surround nursing, issues that play out in clinical moral concerns of the nurse. Baer et al note that

> Identity questions about who the nurse is, what constitutes nursing responsibilities, and what society and the profession can or should expect from nurses are governed by nurses' ever-present desire for power and authority over their work, a yearning that marks every human endeavor. Changing hierarchies within nursing and the social forces that determine nursing's position in society reflect ongoing debates about how the system operates, who changes it, upon whose authority such change is predicated, and ultimately who brings proposed changes to fruition. Nursing's expanding knowledge base raises further questions about what constitutes nursing knowledge, who owns it, who exercises it, and finally who benefits from it? [30].

Over the past century and a quarter much has changed for nursing though for each generation it may seem not enough. And yet, women acquired the vote and the end of coverture and gained some authority in society, including gaining access to legislative positions; nursing education has been standardized in terms of accreditation, and has moved from hospitals into colleges and universities; nursing students have ceased to be the hospital staff, and nursing faculty shifted from physicians to nurses; graduate education has been developed in nursing, including the doctorate; nursing wages have become salaries and have increased to livable income and nurses were brought under labor law; laws have grown to undergird advanced nursing practice; nursing has a seat at the table of commissions, boards, and policy bodies, even if only in token; healthcare teams have become more collaborative and cooperative, even where it remains to be fully realized; the proportion of women in medicine and men in nursing have increased, but not equalized; the National Institute of Nursing Research has been founded as a division of the National Institutes of Health (https://www.ninr.nih.gov); federal funding for nursing education and research has increased; nursing research and evidence-based practice have grown; ethics education in nursing has advanced in its rigor and analysis, though it is not fully and uniformly implemented in curricula by faculty with formal ethics competence; and more. Gains have been made; gains are yet to be made. For whatever gains are yet to be made, nursing is, nevertheless, in a stronger position within the healthcare community of moral discourse, and in a firmer position from which to navigate moral distress with strength, rigor, and vigor.

3.1.2   Social Work Perspective: Moral Distress

Sophia Fantus

3.1.2.1 Introduction
As members of multidisciplinary healthcare teams, hospital social workers are often held accountable in assuming responsibilities that are, at times, outside their clinical scope and professional expertise and skill-set. For instance, tasks may range from administrative duties and discharge planning, to therapeutic support and case management [31, 32]. Moreover, social workers uphold competencies in problem-solving, patient advocacy, as well as issues pertaining to social justice and ethical practice. As a result, social workers may be called upon during times of increased stress and ethical conflict to assist in resolving disputes between patients, families and healthcare teams [33–35]. Accordingly, social workers' occupational roles across hospital settings may trigger reactions of moral distress, discerned as individual responses to resolved ethical dilemmas that have compromised one's moral integrity and professional code of ethics [36, 37]. However, limited research has investigated the experiences of moral distress in social work [35]. A lack of theoretical and empirical scholarship has created difficulty in naming, addressing and subsequently mitigating social workers' moral distress. However, explicating sources of moral distress in social work is imperative to inform practice, education and research. Importantly, this commentary will elucidate how moral distress may transpire for hospital social workers.

3.1.2.2 Limitations in Empirical Scholarly Work
The concept of moral distress has primarily been explored in nursing and medicine to identify how withdrawal or administration of treatment, end-of-life care and patient treatment decisions may trigger reactions of moral distress [38, 39]. Findings have shown that nurses report high incidences of moral distress, often associated with: (1) the administration of aggressive and/or futile treatment; (2) working conditions, including staff shortages, budgetary concerns and increased workloads; (3) power differentials and hierarchies within the healthcare system and across healthcare professions; and (4) issues of self-doubt, fear and an inability to complete tasks [25, 40–42]. Thus, a broad range of individual, interpersonal, and systemic factors trigger experiences of moral distress, and lead to moral compromise and value conflict. Overall, moral distress may have deleterious consequences on the quality of patient care, effective job performance, as well as one's satisfaction and engagement with work [43–46].

More recently, comparative scholarly work has started to emerge investigating non-direct vs. direct care professionals' experiences of moral distress [47–50].

Notwithstanding the importance of such research, social work participants often comprise smaller samples when compared to participants working in nursing or medicine. Consequently, this engenders obstacles in performing cross-discipline analyses; the ability to recognize nuanced discrepancies between direct care practitioners and allied health professionals is difficult. Understanding professional differences is important to identify, address, and mitigate moral distress across disciplines.

In both empirical and theoretical social work scholarship, moral distress has not yet been adequately differentiated from other deleterious experiences. For instance, *burnout* and *occupational stress* are ubiquitous terms in social work practice and may have similar root causes as moral distress (including workloads, relationship conflict and resource constraints). Yet, burnout and occupational stress do not necessarily arise from morally comprising and ethically conflictual situations; rather, they are responses to general occupational constraints and pressures rather than inherent value conflicts [51–54]. Moreover, concepts such as *disjuncture, ethics-related stress* and *professional dissonance* have started to emerge in scholarship to elucidate issues that, at the forefront, seem quite comparable to moral distress. However, the inconsistent and arbitrary language used to describe these experiences subsequently hinders: (1) the ability to effectively address and identify deleterious consequences arising from moral distress; and (2) the synthesis of empirical scholarly research to establish evidence-based practices to mitigate experiences of moral distress.

DiFranks's study [55] investigated social workers' ($n = 206$) *disjunctive distress*, when beliefs in the professional code of ethics are discordant from (and not reflected in) behavior. Survey items included: (1) there have been times when I have had to compromise my professional integrity in my job settings; (2) I have experienced frustration because managed care and bureaucratic constraints often require termination before the client has been able to change; (3) l feel stress at work because I am not always able to help people in need with their personal problems and help them improve larger social issues; (4) I experience stress because of the conflict between my individual clients' interests and my agency's interests; and (5) I feel increased stress because, at times, my professional integrity has been compromised by practice realities. Although these items may support moral distress in social work, what remains unknown is whether these instances are the result of value conflicts from ethical dilemmas.

Similarly, a study conducted in Scotland looked at criminal justice social workers' ($n = 100$) *ethical stress*: comprising both disjunctive distress and ontological guilt [56]. The concept of ontological guilt refers to the accrual of regret that develops from acting in conflict with one's individual values and ethics; social workers may feel that they are not always able to help their clients. Moreover, Taylor [57] investigated *professional dissonance*, the discomfort that stems from conflict between professional values and expected occupational tasks. In consequence, ethics-related terminology presented in social work scholarship has elicited confusion in how to discern the concept of moral distress from other ethically challenging occurrences. This has important implications for social work practice, policy and education.

3.1.2.3 Moral Distress in Social Work

Although ubiquitous in medicine and nursing scholarship, the concept of moral distress has been inadequately explored in social work. As social workers' roles do not involve concrete decision-making with direct medical interventions and treatments, moral distress may transpire in entirely unique ways. Interpersonal relationships, advocacy, problem-solving and mediation are imperative skills that are a critical part of a social worker's hospital role. For instance, in a recent study on moral distress, ancillary staff ($n = 7$; 24%) (including four social workers, two chaplains, and one case manager) reported that moral distress transpired from "family-to-family discordance" more frequently than physicians ($n = 6$; 21%) and nurses ($n = 16$; 55%). Participants described that "working through family dynamics and psychosocial-spiritual barriers, occasioned frequent interactions with family members and patients that could create moral distress" [47, p. 826]. Similarly, Ulrich et al. [50] looked at ethics-related stress (a negative outcome of moral distress) by showing how social workers and nurses ($n = 1215$) reported feeling powerless when dealing with ethical issues, overwhelmed at ethical decision-making and increased job difficulty on account of ethical issues.

Few studies have exclusively investigated (and labeled) moral distress in social work. A recent study in Finland [49] specifically examined reactive moral distress. Reactive moral distress, or moral residue, results from recurrent moral distress that intensifies and escalates over time [25, 38]. This study assessed: (1) work-related mental well-being; (2) acting in accordance or in conflict with professional values; and (3) encountering insufficient resources, such as budget constraints and unmanageable workloads. Among respondents ($n = 817$), 77% felt that they were often unable to do their work as well as they would like, 36% felt that they were often forced to work in a way that conflicted with their professional values, and 18% experienced impaired work-related mental well-being at least a few times a week [49]. The authors suggest that participants who experienced less moral distress reported enthusiasm, inspiration, pride and resilience in their work more than those who experienced greater moral distress. However, a limitation with this study is that not all items/situations presented were indicative of moral distress.

Additionally, an Israeli-based study looked at moral distress among 216 social workers in long-term care facilities [58]. The authors administered a survey to social workers with items, such as: (1) I acted in a way which has been in contradiction to my professional beliefs due to pressures by the institution's management; (2) I confronted the staff when I perceived their behavior as being in contradiction with the best interests of the residents; and (3) there were situations in which I felt that my professional obligation to the residents was in contradiction with the financial interest of the institution [58]. Although findings reported low levels of moral distress among participants, this is perhaps indicative of the relevancy of survey items. Without utilizing a standardized validated measure, such as the Moral Distress Scale [59] or the Moral Distress Scale-Revised [60], it is important to consider how moral distress was conceptualized among social workers.

3.1.2.4 Root Causes of Moral Distress in Social Work

The root causes of moral distress for hospital social workers may operate across four distinct levels: (1) interpersonal clinical interactions, (2) working conditions, (3) power differentials, and (4) professional competencies, skills, and ethics [61].

Clinical interactions may be associated with end-of-life care and advance care planning discussions [62–64]. Social workers often uphold responsibility to advocate for patients and families if a patient refuses treatment, withdraws from futile care, or dismisses the healthcare team's advice and/or recommendations. Although conflict may ensue across the multidisciplinary team and the social worker may not agree with the resolved course of action, the social worker's responsibility is to advocate for the patient's wishes. The social worker may thus be in a position of disagreement with either the medical team or the patient's choice in treatment. Consequently, social workers "not only carry responsibility for moral decision making exercised at the higher levels of the public administration, but they also carry responsibility for their own moral decision making on the individual level, in their face-to-face encounters with their clients" [49, p. 88].

Working conditions can include budgetary constraints, staff shortages, and unmanageable workloads [65–67]. Funding shortages may result in discharge planning that may trigger moral distress [45, 68]. Organizational constraints may lead social workers to carry out discharge plans that conflict with their professional and personal ethics. This may intensify when the patient does not have the proper supports, finances or networks in place after discharge. Thus, when a social worker knows that the patient requires additional assistance, support and routine care that may not be adequately managed or implemented, and yet the hospital has required her to be discharged, this may result in moral distress.

Power differentials may reflect limited job autonomy and hierarchical power imbalances that place social workers in sometimes ethically compromising situations. Clinical social workers are members of multidisciplinary teams, yet they often lack control and autonomous decision-making in their workloads, patient care, and resource allocation [69]. Social workers may be hesitant or uncertain in how to confront occupational conflict, perhaps owing to disempowerment and shame [70, 71], a lack of supervision [56, 72, 73], and the overwhelmingly female-dominated profession of social work [74]. When social workers continuously feel unable to challenge the resolved ethical decision (often due to power differences), they become silent; social workers describe themselves as being omitted from ethical decision-making and patient treatment plans [75]. Poor collegial support, inadequate supervision, and a lack of inclusivity and collaboration with social workers may foster moral distress.

Professional Competencies, Skills, and Ethics can result in the manifestation of moral distress when social workers either feel as though they do not have the competencies or skills to perform their occupational role or there is conflict between their job performance and their professional code of ethics [55, 76]. This may be a result of role conflict (conflicting demands of their job) or role ambiguity (lack of clarity in expectations). As colleagues may not completely understand the range of social work skills and competencies, this often leaves social workers responsible for tasks that are not in their job description. Role ambiguity and conflict may influence social workers' *self-perceived competence*, the "subjective evaluation of the person's skills and abilities to perform well" [75]. In a study among 591 social workers in the state of New York, participants' higher levels of self-perceived competence were associated with lower levels of emotional exhaustion and symptoms associated with burnout. The author posits that lower levels of self-perceived competence may impact one's ability to effectively resolve and react to one's job performance [75]. A lack of self-perceived competence may have consequences on workplace relationships and the perceived ethical climate, and in turn important consequences on how moral distress transpires across social work professionals.

3.1.2.5 Implications for Social Work Education and Practice

Addressing the concept of moral distress in social work has important implications for both education and practice. Social work educators must learn to advance ethics coursework through distinguishing between ethics-related terminology, and addressing the manifestations of moral distress in social work. Identifying moral distress can perhaps help ready future social work practitioners to recognize and name these experiences, and address ways in which to mitigate deleterious consequences in their professional context. Discussing moral distress may also support multidisciplinary dialogue and conversation [35]. Utilizing common terminology may assist social work practitioners to seek supervision and support from other healthcare practitioners and to find common methods to prevent and resolve such conflict.

Future research is necessary to empirically understand how moral distress may transpire among social workers. Pilot studies can lead to the establishment of measurement scales to explicate specific items relevant to social workers' duties and responsibilities that may lead to moral distress. Furthermore, such pilot studies and validated measures can help further understanding of how moral distress differs from disjuncture, burnout, occupational stress and professional dissonance, and seek evidence-based practices to identify, discuss, and process the experience of moral distress.

3.2   Part 2: Healthcare Professional Perspectives

3.2.1   A Source of Moral Distress: The Corporatization of Medicine

Joseph J. Fins

Over the past decade, I have seen far less moral courage and the sort of professional autonomy that allows doctors speak out against what they may perceive as wrong or improper behavior. I focus on doctors, not to be physician-centric, but rather because there has been—in my view—a pronounced change in physician behavior in the three decades I have been in practice.

I have detected a decline in the sort of independence that docs were known for and that attracted many to the profession. This begs for an explanation and has implications for professional autonomy and its obverse when physicians feel a sense of moral entrapment and it is impossible to speak up or out.

This is a relatively recent change. There was a time when physicians prided themselves on their ability to self-regulate—whether they did so or not is another topic for another time. Physicians were emboldened by a sense of professional autonomy and discretion that allowed them to set their own moral compass and proceed in the direction they thought best and right. A downside of this hegemony was paternalism that fortuitously has been countered by the emergence of the patient's voice. But there was an upside to this sense of professionalism, the ability to express one's views as a physician. And with these articulations came a sense of empowerment that comes with having one's opinions heard, respected, and acted upon.

This professional prerogative has been eviscerated by many factors but one key sociological force has been the emergence of corporate structures of care that have tempered the power of the individual doctor and led to conformity and complacency. This becomes obvious if we contrast the private practitioner of yore with the hospitalist of today. The doctor in private practice a few decades ago was generally self-employed. Still a predominantly male profession, he was paid directly by his patients or their insurance companies. He was neither an employee of a managed care company nor the hospital and thus was independent of any financial pressures that they might exert. Indeed, the private practitioner in prior decades had power *over* the hospital as he was a source of patient revenue because he directed patients to one hospital or another depending upon where he chose to admit patients. If he was maltreated or censored in any way he could retaliate by redirecting this revenue stream and sending his patients elsewhere. This economic clout conferred power and the requisite independence which is sometimes needed to speak up.

Contrast this now quaint model with the modern hospitalist who is a full time employee of the hospital. His patients are assigned, and his patient load set, at the discretion of the hospital where he works. His wages are fixed, sometimes sweetened by a year-end bonus which is dependent upon efficiency and adherence to length of stay metrics. Any effort to counter hospital policies could imperil one's standing at the institution and potentially compromise one's employment, notwithstanding platitudinous standards about professionalism and accreditation standards.

These constraints are further compounded by the hierarchical nature of healthcare institutions with the decline of powerful departmental chairs and the rise of central administrators who control budgets and their chairs through the power of the purse. This diminution of professional sources of authority, a zero-sum game due to the rise of corporate power leads to further marginalization of clinicians who heretofore would air their grievances with their chairs.

While I am sure these compounding variables depend on the nature of each institution's leadership structure, there seems to be less recourse to professional channels of appeal thereby leading to moral distress and professional estrangement. Institutions that are mission driven by religious or secular attestations of purpose, in my experience, seem less prone to these distortions of professionalism.

In the aggregate, conflicts of interest and the corporatization of practice can constrain patient advocacy and lead practitioners to feel torn between their obligations to their own families and those who are entrusted to their clinical care. This is a challenge for professionalism and can lead to moral distress. Increasingly, clinical ethicists are being called upon to use whatever institutional moral authority they have to provide a remedy and counterweight to these forces. Our advocacy echoes responses that may no longer be available to individual practitioners. Hopefully our efforts can do more than respond in individual cases and help corporate leaders of medicine appreciate that professionalism, as seen in the requisite autonomy of practitioners to be moral agents, is the healthcare system's greatest asset. If we lose that element of care, the loss will be priceless.

3.2.2   Moral Distress: A Psychiatrist Perspective

Michelle Joy

As I think of my job—my experiences as a psychiatric fellow, clinician, and forensic evaluator—I am first grateful. My work gives me a true sense of contentment and of appreciation, and I am keenly aware that I am very lucky to enjoy the work that I do. I hold the patients and evaluees that I see in high esteem. I intrinsically respect them. But beneath these interpersonal interactions, there exists a darker reality. It is the structure of society, the systems of care and incarceration. It is hierarchy and inequality. It can be confusing, difficult, and altogether distressing.

Working in community and forensic systems of care, you get the sense that psychiatry becomes an attempt to hold together the underfunded, under resourced, underdeveloped parts of society. And I don't mean the psychiatric care itself. You run abreast of poverty, food insecurity, homelessness, lack of access to healthcare, and limited education. And sometimes people are coming to you not because of the stresses of those situations—yes, that too—but literally and directly because of their needs. Psychiatry can become a route of access to social security disability payments. Psychiatry can be "three hots and a cot"—a colloquialism for food and shelter provided by the hospital (or even jail).

Frustration lies not with the individuals but with the dance itself. Suicidal becomes a code word for "I need something and can't be turned away." But how can you blame the person—subjecting themselves to intrusive questions, often medications, long waits of hours to days, loss of autonomy and privacy and freedom in admission to a psychiatric hospital…. The distress comes with knowing that this won't—and can't—be fixed with a pill. That there is no prescription a doctor can write to change society. But while wishing for more, for this person, for everyone, you try. You try to inspire hope, to validate struggles, to empathize with difficulties, and always to respect the individual.

And it can be hard. The dance goes on all hours of night and day; you do this first thing in the morning, at 3 in the morning, late at night. You can be yelled at, called names, or worse. You hope to avoid fecal smearing and assaults at all costs. But you remember that a lot of this exists beyond diagnosis. It is the desperate cry of people in need, people without, people living in a rich and glamorous but wholly unequal society.

And too there is the trauma. Horrific stories of rape and torture, physical abuse, molestation during which you do your best to be present and honor survival.

And too there is the stigma. You know it from living life—see it on social media, hear it at dinner parties, receive it as mental health practitioners (always a joke to be had) or as people with diagnoses ourselves. You call and receive consults in which psychiatric comes to mean difficult, annoying, someone others don't want to deal with. And in quiet moments you realize that stigma is probably keeping many, many people from even making it to your door.

Also the discrimination. In a racist society, stories abound. A black man arrives late at night to discuss nightmares and fear after racial tensions in the military, his fear of white men, his fear he will retaliate. You realize there are no black providers to hold this space with him, and you try your best. A black child in a detention facility won't speak with you—despite your assignment to see if his case can be helped. You learn he is willing to speak with a black psychiatrist; you lament that no one is available. He falls down the roster. The transgender patient hears dead names, inappropriate pronouns, and "but have you had surgery?" all too often.

The system itself twists and turns and disappears behind layers of complexity. As a provider or a patient you try to grasp the complexities of insurance, referrals, prior authorizations, copayments, deductibles, sliding scales, waiting lists, appointment scheduling, refills, and more. And then you imagine attempting to navigate this without a phone. Without a home. Without money. Speaking another language. Lacking motivation. Distracted by hearing voices. With no one to help. Or care. In the forensic system the playing field is populated with attorneys, judges, plea bargains, evidence, probation, rights, and waiting. How to navigate is again the question.

And even within the services themselves, they glimmer then dart, a disappearing school of fish in a dark and infinite sea. Insurance will only pay for a short course of therapy. The dialectical behavioral therapy program won't take someone with an addiction. The trauma program will not accept someone who is suicidal. The early psychosis program only sees people within months of symptom onset—too late. The therapists regarded as the best charge hundreds of dollars per hour. The psychiatric facilities that treat complex medical problems close down. The medication has unbearable side effects. A treating provider has left the training program, the area, the field.

There are things we are forced—but are we?—to do that keep us up. Decisions we make. Protocols we follow. Involuntary commitment: did I save a life or traumatize someone, ensuring they will never again seek services? Malingering: was he really fabricating a story or was I just tired, frustrated, and resentful? "But the hijab must come off, it's a psychiatric emergency room, and we can't have anything someone could hang herself with." Do I have a suspicion of abuse, does this family require child protective services, or will that just be another stress and possible trauma? Declaring capacity to refuse treatment might mean capacity to accept death. Competency to stand trial means going forth with all the possibilities a guilty verdict might entail; not competent is waiting longer in jail. Will providing a diagnosis impair or empower? Should I suggest medications or therapy, neither or both? Because as much as the roles we fill ask for, insist upon, prediction, we are not fortune tellers or lie detectors. We hold no magical abilities and often operate with limited time and information. There are few lab tests or scans as ours is the world of words and stories.

But that world is a special one. It is a world of which many don't know hidden behind locked doors and privacy protections. But let it be known that despite its frustrations and flaws, it is a wonderful space to inhabit. We spend our days holding narratives, emotions, thoughts, and behaviors of most personal natures. I thank those individuals who share with me and hope that our system and society can work to increasingly improve the ways in which we can help those in need.

3.2.3 Physicians' Experiences of Moral Distress and Burnout

Katherine E. Kruse and Alyssa M. Burgart
Some physicians are unfamiliar with the term "moral distress," but upon hearing a description of the concept, they invariably realize they have personally endured moral distress or witnessed its aftermath. In the past decade, nursing literature has taken a deep dive into the study of moral distress, while medical literature has focused on the closely related issue of burnout [77]. Recently, the disciplines have converged and we have seen increased work on the interrelatedness of the concepts and of our professional experiences. Such studies bring to light the connection among moral distress, burnout, and depression, and their correlates: individual resilience, institutional moral climates and moral community. Nurses, physicians, social workers, other bedside providers, as well as hospital administrators, may experience moral distress [78]. The physician's role in modern healthcare carries specific professional expectations which are distinct from other roles. Our facets of responsibility are defined on several fronts: individual (patient expectations for a physician's care), societal (promotion of the public good), logistical (medical licensing requirements), legal (risk for malpractice claims), and personal (a physician's desire to be perceived by one's self and others as a "good doctor"). We may experience moral distress across the spectrum of professional life: the care of critically ill patients, working with challenging families, conflict with administrators, limited access to services and resources for patients, legal matters, policy constraints, among others. The effects of such stress do not stop with individual clinicians, and are implicated in harms to patients, such as medical errors [79].

Physician professional identity formation, anchored in the societal and professional expectations unique to our brand of medicine, leads to development of an exceptional sense of personal responsibility for our patients [80, 81]. By nature of our vocation we are held to a higher standard than non-medical professions, and patients insist we remain unblemished to gain and maintain their trust. Our contract with society expects that we serve as healers, guarantee our competence, be altruistic, act morally and with integrity, promote the public good, and be both transparent and accountable [82]. Armed with medical degrees and years of specialization training, the public requires physicians to be top notch diagnosticians and clinicians, but also scientists, teachers, and role models. We are expected to be both human (akin to our patients) and simultaneously superhuman heroes (capable of saving lives). The public barometer of success is no longer measured by accurate diagnoses or lives saved, but consumer ratings where those disgruntled with healthcare tend to be the most vocal.

Unmitigated duty to, and personal responsibility for, one's patient are non-negotiable elements of physician practice. It is a duty and a privilege to care for the ill and dying, but can be burdensome as well. With this level of responsibility, even when we support team-based practice and a shared decision-making model, physicians often see ourselves as carrying much of the responsibility for ensuring individual patient outcomes. For many of us, the work for a patient doesn't end when we leave the hospital or clinic, as our patients remain on our minds throughout the day, sometimes even appearing in our dreams. This quest to serve each patient may become all-consuming. When coupled with the administrative tasks associated with practice, we are known for chipping away at time for personal care, making the achievement of an already nebulous work-life balance impossible. This trajectory sets physicians at risk for losing hold of the deeply rewarding and meaningful aspects of professional life.

Physicians frequently operate under the umbrella of a larger organization with its own priorities and obligations, which may conflict with the medical goals of individual patients or the best intentions of staff. When organizational values are non-congruent with those of physicians, morally distressing conflicts arise. Employers may mandate an unrealistic number of patients to be seen in one's clinic, leaving physicians to balance the fallout of one complex patient's needed care, leading to a waiting room full of irritated patients whose appointment times have long since passed. Physicians may also be expected to maintain Press Ganey patient satisfaction scores, which are themselves correlated with patient perception of sufficient time spent with the physician [83]. While in training, physicians anticipate spending their days providing direct patient care, but actually spend almost twice the amount time doing clerical work, and even more hours at home to complete it [84]. Some moral distress is unavoidable in our line of work and the risks for burnout will never be eradicated. Successful organizations acknowledge this reality and attentively foster a strong moral climate, nurture resilience, and balance demands on physicians so that we can forge ahead, rather than become disenchanted with the practice of medicine.

For some physicians, the combination of lofty expectations, a deeply ingrained professional integrity, low resilience, and untenable professional/institutional expectations create the perfect breeding ground for moral distress, burnout, and depression. Physicians who find and appreciate the deep meaning in their work are far less likely to experience moral distress and burnout. However, medical training does not require us to be emotionally healthy people armed with good coping abilities, resiliency, moral sensitivity, and ethical discernment skills, nor are such skills specifically nurtured in the arduous process of becoming a physician. Many physicians find themselves well trained in medicine, but woefully underprepared emotionally for its stressors. At the core of our calling to be doctors, sometimes lies both our greatest strength and the seed of our undoing. Our drive to be the very best clinicians leads us to spend long hours caring for patients, voluntarily cutting into time with our families and personal interests. Self-care is so easy to cut from one's day, and we can end up emotionally and physically drained, hindering our moral sensitivity and perspective. Unregulated moral distress may lead to moral outrage, burnout, and acute secondary stress [85]. While it is tempting to see this as a personal problem, moral distress and burnout are associated with increased rates of medical errors, meaning that patients suffer as well. Some physicians' professional quality of life is so impacted that they leave practice altogether [86]. Physician burnout and depression are strongly correlated [87, 88], and some argue are one in the same [89]. Though it may initially sound dramatic, the string that connects these phenomena may be life threatening. An estimated 300–400 physicians commit suicide annually. This tragedy is not well understood, but is believed to be due to a combination of burnout and untreated mental illness [90].

3.2.3.1 Physician Narratives

As physicians and clinical ethicists, we navigate the murky water where clinical care and ethics converge. Sharing our professions' stories of moral distress is a wonderful way to open this important dialog. By acknowledging the difficult aspects of our work, we begin to prepare ourselves and our fellow clinicians to move past survival and create space to thrive in our work. To highlight moral distress in the clinical arena, our colleagues graciously shared their experiences:

Legal Rights in Organ Donation: Directed Donation

A heart failure specialist considers fair practices of organ transplant allocation.

> I know it's [the family's] right [to give the organs to a specific person], but it feels really wrong. For the patient that gets the heart, if it's a good match, it's great, but it means someone who is really sick and may be top of the list, won't get it… and that person might die because they weren't lucky enough to have a friend die. It sounds sick, to say that… The UNOS system is supposed to make it so we don't have to be involved in the details of the donor. When the organ is directed, suddenly, the donor is much, much closer… It makes us all really uncomfortable.

Right to Information: International Medical Care

An international disaster relief physician struggled when practices around HIV were abruptly changed.

> Our mission was a 'chronic emergency'… we had a strong presence and had been [in that city] for over five years… They had been testing for HIV and there was actually a way that we could request HIV medications on a case by case basis… But while I was there, we got an order from [the organization] to stop testing all together… we thought, even if we couldn't treat or continue the responsible care, we really felt that the patient had a right to know and definitely had a right to be tested. We were there to provide care. The test was simple and we had the time to do it. But this wasn't a pandemic, like ebola… it wasn't a crisis, so we could take care of every patient that came in front of us. I had a real problem with not being able to do the test for my patients.

Patient–Physician Communication: Gestational Carrier

An obstetrician uncomfortable with limitations placed on her communication with her patient.

> I had a patient who was a gestational carrier for a couple… and because of the agreement with [the surrogacy organization] I was being told that I wasn't allowed to tell my patient about everything that was going on [a severe cardiac defect] with this being growing inside her body… Normally, I would have been able to do that… to talk about it, it's part of the relationship. You just want to take care of the person in front of you… Then [because of the disability] the parents didn't want the baby anymore… and the surrogacy organization basically stopped said 'we're done.' So then I thought, who's in charge now? Who gets the information about this baby? I just wanted to talk to her.

These examples provide a glimpse into physicians' morally distressing experiences. We encourage ongoing effort and focus on morally distressing events, both large and small, occurring in the practice of medicine. No matter the magnitude or flavor of moral distress, all merit respect and consideration [91]. The prevention and treatment of moral distress requires stepping back to examine the deep meaning that drives physicians to choose careers in medicine, the environment of practice, and what it takes to foster and support a morally robust community in which such physicians can thrive.

Moral distress takes many forms and can permeate every aspect of our professional lives. We have come a long way in recognizing moral distress and its connection to burnout as significant problems in medicine. Organizations across the country are making efforts to create better moral climates for all healthcare providers and patients. Acknowledging moral distress head-on, before it can smolder into burnout and depression, is one approach to ensure career longevity for those in the thick of it. When healthcare teams work together to address moral distress and burnout, we can make immense strides towards a more resilient moral community.

3.2.4   A Chaplain's Perspective on Moral Distress

Margaret Lindsey

I first became interested in moral distress when I was working as a chaplain at a suburban Chicago hospital and beginning the coursework for a Doctor of Ministry degree. I attended a conference on perinatal loss, heard a presentation about moral distress, and was hooked. What struck me immediately was how well suited chaplains are to respond to the problem. I had the good fortune, at the time, to be leading a series of seminars about medical ethics for the residents at our hospital, and I began to wonder if they shared the experience that I had just heard described as a nursing issue. I decided to make that question the focus of my doctoral research.

Why are chaplains so well suited to respond to the problem of moral distress? As I understand it, moral distress is a form of suffering, specifically spiritual, emotional, and moral suffering. It is a crisis of identity for the provider which threatens his or her sense of self as a moral being. It is a threat to the provider's integrity which may lead to a diminished sense of purpose and meaning in his or her work, and often results not only in a loss of job satisfaction but in a painful sense of having betrayed oneself and one's deepest convictions. Chaplains, as members of the clergy, have a fundamental responsibility to attend to the moral lives of those who are in their care. At the same time, it is our basic purpose, as chaplains, to ease spiritual suffering. Moral distress is a crisis that demands the fulfillment of that responsibility and of that purpose.

My research convinced me that moral distress was a universal experience among the medical residents with whom I worked. My conversations about the topic in our hospital's ethics committee eventuated in several other presentations to both nurses and medical staff, where I repeatedly heard the same thing. "Yes, that's my experience! I just didn't know what to call it. I still remember what happened with a patient, years ago. It was awful. Let me tell you about it." Sometimes there were tears. The pain lingered, and the moral residue clung. I discovered how common the experience was. My next challenge was to figure out what to do about it.

Most chaplains consider it their responsibility to care for hospital staff as well as patients and their families. We are trained to be good listeners and to facilitate healing conversations. As members of the clergy, we expect, and are expected, to keep confidences. In most hospitals, chaplains stand apart from administration and management and so are able to provide a safe haven for fellow staff members to discuss personal concerns, such as moral distress. Chaplains can address the issue, first of all, by providing education that describes and names the problem and by offering safe opportunities for providers to tell their stories.

Chaplains can also address the issue by supporting the efforts of providers who decide to work for change. As the American Association of Critical Care Nurses position statement, "The Four A's to Rise Above Moral Distress," suggests, the most adaptive response to moral distress may be to take action, but the risks and benefits of that action must be carefully considered. Chaplains are trained to facilitate decision-making, and so can offer assistance and support as providers assess the situation, evaluate their options, and determine their response. Chaplains can, and should, provide ongoing emotional and moral support as steps are taken toward change. In some hospitals, particularly faith-based ones where chaplains are seen as moral leaders, they may be well positioned to advocate for providers who work for change and justice within the system.

Many chaplains are trained in medical ethics, or serve on ethics committees, and may be able to offer insights from that training as providers grapple with moral distress. Most chaplains, as members of the clergy, have some basic education in ethics and some facility in analyzing ethical problems. Chaplains may, at times, address the issue by offering basic ethics education and proposing various models of ethical decision-making.

Much of my thinking on the topic, however, has been theological, and that, of course, is the unique perspective that a chaplain brings. For me, it's all about vocation. From my point of view, the most important question for a provider who is experiencing moral distress is ultimately "What is God calling you to do?" Of course not all chaplains are Protestant ministers, as I am, and not all providers are Christian. A more universal question might be "Who are you meant to be?" or "Who do you want to be?" or "Why did you choose this work?" Whatever source of motivation one appeals to, there is great power and hope in the recollection of that motivation, and in working one's way through the quagmire of moral distress in order to reclaim what one once held dear. Although it seems a cruel twist of fate that those who are most sensitive about moral problems are also most vulnerable to moral distress, it is also the case that those who successfully wade through the muck emerge stronger. It takes courage and commitment, but is well worth the effort. There is much to celebrate if, in the end, the right thing is done for the patient and the provider's sense of self is restored. From this chaplain's perspective, moral distress is not just a painful problem, but a tremendous opportunity for spiritual, emotional, and moral growth.

Studies are beginning to show that providers from a wide variety of disciplines experience moral distress. In a 2013 study, Susan Houston and her colleagues at Baylor demonstrated its occurrence among a wide variety of healthcare professionals, including chaplains [48]. The chaplains in their study reported a high degree of intensity in their experience of moral distress, and a tendency to be most distressed by patient care situations that raised issues of social justice. It stands to reason that chaplains, who are charged with a particular responsibility for the moral well-being of others, would be acutely affected when their own moral integrity is threatened. It stands to reason, too, that chaplains, who are not just spiritual and pastoral caregivers but also religious leaders, would have a heightened sensitivity to issues of social justice and a consequent sense of responsibility. We chaplains are challenged by the moral distress we ourselves experience, just as our colleagues in other disciplines are. Will we notice it, learn from it, and grow?

My purpose in working with the residents at our hospital was to provide better pastoral care for them. As I became aware of this particular problem, I worked to find ways to ease the suffering they experienced, and to encourage their growth as individuals. That might be enough, by itself. But chaplains and other pastoral caregivers are increasingly aware of the systemic implications of our work, and my exploration of this topic has convinced me that our response to it, both as individuals and as institutions, has the potential to have a far greater impact. Might it not be that happy, spiritually healthy providers provide better care for their patients? What if more healthcare providers felt well-equipped to navigate the shoals of moral distress and work for positive change? Could it be that their efforts would lead to much needed improvements in our healthcare system and a better healthcare environment for us all? I'm betting on yes. As painful as the experience of moral distress can be, the opportunity it presents gives me hope.

3.2.5   Moral Distress in Pediatric Nursing and Research

Kim Mooney-Doyle

> To cure sometimes; to relieve often; to comfort always.

As I embarked in a career as a pediatric oncology nurse, I knew that suffering and death would be part of the journey. I couldn't control that. I cannot control if a beloved child develops cancer, how they respond to treatment, whether they relapse, and if their disease causes their death. What I can control, however, is how I provide care, how I teach others to provide care in this context, and the research questions I choose to investigate that may elucidate sources of child and family suffering, strength, and distress. I can work to minimize suffering and help children and families process life-threatening situations and decide as a family how they want to live life and how they want to die. I can try my best to be a source of sanctuary to children and families who live with life-threatening illnesses.

Without a doubt, one of the most life-giving aspects of working in pediatric healthcare is the relationships formed with children and their families. It is painful and confusing to see a child's symptoms mismanaged as they become more ill and death draws closer. It is painful to see families and clinicians have discordant views about prognosis and potential for cure. The pain that sticks with me, though, is the pain of seeing families not get what they need from the healthcare system or healthcare providers. This lack of support takes many forms: the single parent who is told that she cannot stay at the bedside of her sick child with the younger healthy sibling; the teenage girl and her mother whose complaints are blown off because they are deemed high-maintenance (and the child ended up in the intensive care unit); the mother who has one sick child in the hospital and other children at home and perceives that she is judged and "feels treated like shit" when she comes to the unit to see her child because she can't be at the child's bedside constantly; or parents who are judged by healthcare providers as "in denial" or "uninformed" for decisions they make about their child's care when the healthcare providers have incomplete information about the clinical or family perspective. Indeed, my moral distress as a pediatric nurse and researcher is rooted in the limitations of support provided to families; it is the lack of recognition of family moral distress.

Much of the recent literature on moral distress in pediatric healthcare providers describes the sources of this distress and how it varies among healthcare providers [92]. In addition, other literature points to ways in which moral distress can be minimized through innovative, interdisciplinary communication [93] or through instituting high-quality palliative care for children in a given unit [94], acknowledging the bidirectional relationships among pediatric healthcare providers, children, and their families. We influence the children and families we care for, and they influence us in return. Indeed, various disciplines experience moral distress for a multitude of reasons, situated in their given professional context, yet there are common themes throughout: providing care that seems futile or that causes harm; poor team communication; and discord between family and staff appraisal of a child's clinical situation. Yet, healthcare providers in these studies less frequently express concern about the way children and families are treated in the healthcare system or how children and families live with life-threatening illness and how they survive as a family unit. Also, parents, children, and other family members have had limited opportunities in the literature to express their own experiences of moral distress. It seems as though our concern about moral distress is more about us as healthcare providers and less about children and families. It seems as though we have forgotten that we get to leave the four walls of the intensive care unit or the oncology unit and, often, return to our healthy loved ones. Yet families are trapped in that existence, may not be able to fulfill the expectations they have established for themselves as parents, and may end up leaving the hospital without their child.

One of the greatest sources of moral distress described by pediatric healthcare providers is when a family desires and requests increasingly aggressive care, when the healthcare provider does not perceive it will be beneficial for the child. Yet, what rarely seems to be part of the conversation is how we, as pediatric healthcare providers, have a hand in creating this distressing situation. There is ample evidence in the pediatric oncology literature, for example, that parents and oncologists frequently have different prognostic expectations for a child's advanced cancer [95], that pediatric healthcare providers struggle with sensitive and difficult conversations about transitioning care in life-threatening illness [96] and that they fear diminishing hope and causing distress [97, 98], that there are communication gaps between healthcare providers and parents [99], and that parents want honest, clear information from healthcare providers delivered in a caring way [100]. Thus, illuminating the relationship between communication and moral distress for healthcare providers and families may be an important way to address the experiences of moral distress in pediatrics and mitigate its effects.

Another risk of these gaps in communication that may contribute to moral distress or result from moral distress is "othering" of parents and children who make decisions with which we do not agree. "Othering" is a process in which "a particular social group becomes defined or characterized in contrast to the dominant social group, usually with hierarchal undertones" [101]. As further described by Whitehead [101], "othering" allows those who perceive a wrong to "engage in a meaningful, therapeutic exercise that shifts their role from that of victim to that of judge. Doing so restores control in a situation that they are experiencing as extremely chaotic or senseless. They manage the chaos of their situation by reordering the occurrence of events in their lives, such that they refile themselves in the 'normal' pile that they are used to being a part of" (p.115). Thus, when pediatric healthcare providers feel they are participating in care they do not agree with or perceive as futile, they may perceive the parents making such decisions as "different" than themselves in order to process the situation, but with potentially dire consequences for the relationship. For example, when we, as healthcare providers, declare that we would never subject ourselves or a family member to stem cell transplantation, yet we have never had to make such a decision, we risk creating an artificial separation between ourselves and children/families. When confronted with a life-threatening illness, we might decide differently.

Eliciting sources of moral distress in families can prompt healthcare providers to see a multifaceted picture of family life in pediatric life-threatening illnesses. Understanding parents in the totality of their roles and situated within their given contexts provides windows into their decision-making and "re-goaling" [102] over time. An ecological perspective can elucidate moral distress within the context of pediatric life-threatening illness [103, 104] (Fig. 3.1). This perspective places the child at the center of various environments, nested within one another. Immediately surrounding the child are the parents and siblings and other close, intimate relationships. Surrounding the child and his or her loved ones is the community environment that encompasses school, friends and peers, healthcare systems and providers, place of worship, among other sources of support and service. Surrounding the

Fig. 3.1 Social ecological framework for understanding moral distress in pediatric life-threatening and life-limiting illness

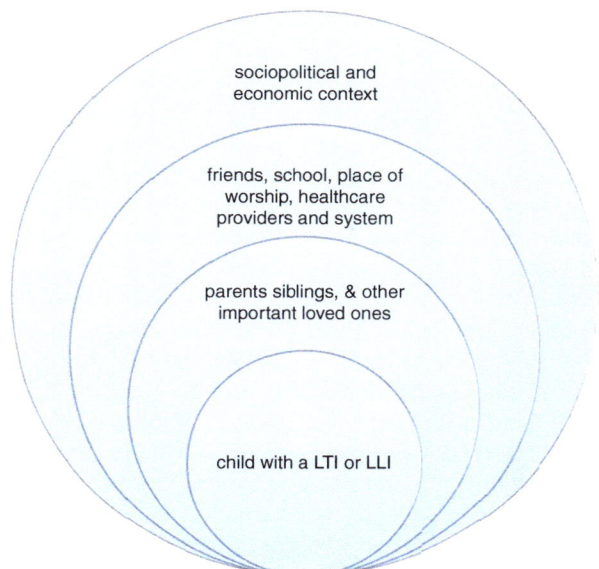

community that envelopes the child and family is the broader system that may not directly interact with the child and family, but influences their well-being. An example of this is the political context that supports legislation to provide concurrent curative and hospice care or family medical leave. Finally, all of these systems are situated within the broader culture that establishes norms and expectations (e.g., gender roles, family roles). These systems influence and are influenced by each other and change over time. Thus, this perspective recognizes that children and their families are the focus of our care and service and that there is a bidirectional relationship between children/families and healthcare providers. Yet, the ecological perspective reminds us that we are but one part, albeit an important (often life-sustaining) one, of a greater world that the child and family inhabits.

Using this perspective to elucidate the experiences of children and families in the context of serious, life-threatening illness, we can appreciate the various sources of stress and strength with which families contend, the meaning parents attribute to their child's illness and their role in being a parent, and the barriers families face trying to accomplish what they deem important [105]. We come to see that in order to feel as though they are "being a good parent to the ill child," [106, 107] parents may believe they should ensure their child has strong spiritual beliefs, may rarely leave the child's bedside for fear of missing a chance to ask the attending physician a question or having the child's needs unattended, or search the country for an open clinical trial. We also come to see the sources of conflict with which parents contend, such as ensuring healthy siblings feel loved and emotionally connected to the parent, which pulls them away from the bedside, financial distress because of lost wages or unanticipated expenses of hospitalization, parents' own emotional or psychological distress [108], or violence within their own homes or communities. Thus, understanding the complex environments that families traverse demonstrates their own potential sources of moral distress and provides insights into behavior and decisions pediatric healthcare providers find challenging.

When we look beyond the action (or inaction) that has instigated our moral distress to the broader context in which parents or children make such decisions, our moral distress may be tempered because we see the situation from another angle that may challenge our initial moral judgement or provide insights into why parents make such decisions. This is similar to research by Laing et al. [109] in which digital stories by children with cancer and their families contributed to healthcare providers understanding of aspects of the cancer experience that were not discussed in a clinical encounter. Through the video, healthcare providers described diminished barriers between themselves and families; by "losing their healthcare provider role" participants in this study were moved by their common humanity with the children and families and perceived greater ability to connect with them [109]. Examining moral distress from an ecological perspective can unearth factors that influence our perceptions of moral distress; we can flip the microscope from the internal to the external. For example, when we change our focus from the distress and negative feelings we experience because a mother does not stay at the bedside of her sick child to understanding that the woman is living in poverty, has other children, and limited safe social support to care for those other children, our own moral distress may be alleviated. We may still find the situation of severe child and family poverty distressing and we may feel sadness for the involved family members and the child who is ill, but we may not feel a threat to our own integrity.

3.2.6 Pharmacist's Perspective on Moral Distress in Palliative Care: A Narrative

Tanya J. Uritsky
I have been a clinical pharmacy specialist working in a large academic teaching center for nearly seven years. I practice in palliative care, working with patients and families in great distress, facing big decisions, and looking for guidance from some of the best and brightest providers in the country, or even the world. They come here to get "fixed" as they often say. They come here because other places have not been able to meet their needs or make them better, but they heard we can do things that others cannot. Unfortunately, we cannot prevent the inevitable, sometimes we can delay it, but often not without consequence of long or frequent hospitalizations, significant pain and anguish. Although we aim to provide improved quality of life, it sometimes gets lost in the incredible push to preserve life. And I am ok with this, as long as it is informed and decisions are made based on "truths" as best as we know them, values are explored, and plans are clear but frequently revisited.

I have had one too many experiences where patients are told an intervention will "help." I really don't care for the word "help" in the medical world. What does this mean? I was working with a very sick patient who was told the chemotherapy would help him—the understanding of the patient's wife was that it would help him have the chance to walk again, regain some function. The intent of the physician was that it would help preserve his organs in their current weakened and malfunctioning state at best, not improve his quality of life. While some may choose to continue in a weakened and debilitated state, that was not in-line with the values of this patient and his wife. From my position as a palliative care pharmacist, I inserted myself between the patient and the chemotherapy and was able to prevent this misalignment from happening. I explored the family's values and clarified with the physician's intent, which revealed the discrepancy in the plan for more chemotherapy. I then worked with the primary medical team and floor social worker to expeditiously establish comfort care for the patient in a preferred location as his health was rapidly declining. My pharmacist colleagues would have been the ones verifying the chemotherapy for this patient, not necessarily knowing much about the conversations or the values going in to the decision to give this medication to a very sick and dying man. I am empowered to try to sort this out as a member of the palliative care team; the unit pharmacists, however, are generally not so empowered.

This is exemplified in the hospital's transition to a new computer system. The pharmacists did not have access to any of the advanced care planning information despite the fact that it was accessible to other members of the medical team. It is presumed, even at the level of technology developers, that the pharmacist will verify a medication, something as major as chemotherapy, because it has been deemed appropriate on some "higher" level. The pharmacist is the medication specialist, with expertise that ranges from the molecular level through the level of interpreting the clinical impact of medications on patients. Pharmacists are on the front-lines; they do much of the counseling to very sick patients about potentially toxic medications and discuss their worries and concerns. Pharmacists may question the appropriateness of a medication order, but without access to patients' advanced care planning information and goals, the implied message is that our perspective does not matter. To rectify this and demonstrate that our perspectives do indeed matter, I worked to ensure pharmacists throughout the institution have access to this information. Unfortunately, the work continues as I don't know how many pharmacists even know they have access to this information, can use it in their clinical work, or feel empowered to do so.

In a different dimension on the above case, sometimes what is perceived as harmful is actually helpful in a way that is not so obvious to the entire healthcare team. I was involved in a case where the oncologist's idea of "help" was in alignment with the patient and his wife, but other members of the team were very distressed since the man was near the end of his life. The other members of the healthcare team had a difficult time reconciling their own values about what should be done with what the patient and his wife wanted and what the physician ordered. In exploring the wife's values, she needed to feel that she had done everything that could have been done—she needed to make every last effort possible to help her husband. It is distressing that chemotherapy was even offered, but I am certain the oncologist was trying to meet this need. This case demonstrates how it is essential to understand family values and the distress that would have lingered with his wife long after this man's death if just one more thing wasn't done. If the pharmacist knows this, there is less strife around verifying the chemotherapy and more ability to offer consolation to the struggling team.

Along these lines, there is the idea that one is "just the pharmacist." The perceived role of the pharmacist can be limiting—as one who only knows the medications or who counts pills. With more and more clinical pharmacy presence on medical teams and with the robust therapeutics education of pharmacy school and post-graduate training, the pharmacist is poised to provide so much more. It has been my experience that providing support and symptom management for those in distress instills trust and this opens the door to explore patient values. Patients look to pharmacists as a trusted member of the team who is now their advocate. I have been involved in complex psychosocial and ethical situations, have led family meetings, have been at the bedside of a dying patient as a support to the family and the staff—all things that do not fit inside the traditional role of the pharmacist. Pharmacists need to be encouraged to get to know patients and advocate for them based on these interactions.

Then there are the moral considerations around stopping maintenance medications at the end of life or when patients have a life-limiting illness. These are crucial conversations and the emotional and psychological attachment that can be linked to the life-sustaining focus of many medications is often the crux of the challenge. The pharmacist is reliant on the prognostication of the providers as well as on their own experience in helping guide the patients and their families through this process. Having experience under my belt, I am less overwhelmed by these conversations, but pharmacists with less experience in this realm may experience distress around these decisions and conversations. As a result, they may be more likely to avoid these conversations or take a more general approach, leaving room for potential distress amongst themselves, the provider team, the patient and their family. Acknowledging this pivotal role of the pharmacist and offering ongoing education and support are essential to providing quality end-of-life care.

The presence of the pharmacist on the treatment team is strengthening and the role is different from specialty to specialty, and in various settings. It is important to acknowledge the areas of distress that may present themselves as this evolves and bring the pharmacist into the conversation about patient's hopes, dreams, and values.

References

1. Jameton A. Nursing practice: the ethical issues. Englewood Cliffs: Prentice Hall; 1984.
2. Jameton A. A reflection on moral distress in nursing together with a current application of the concept. J Bioeth Inq. 2013;10(3):297–308. https://doi.org/10.1007/s11673-013-9466-3.
3. Storch J, Rodney P, Varcoe C, Pauly B, Starzomski R, Stevenson L, et al. Leadership for ethical policy and practice (LEPP): participatory action project. Nurs Leadersh (Tor Ont). 2009;22(3):68–80.
4. Davis A, Aroskar M. Ethical dilemmas & nursing practice. Saddle River: Appleton Century Crofts; 1978.
5. Stein LI. The doctor-nurse game. Arch Gen Psychiatry. 1967;16(6):699–703.

6. Stein LI. The doctor-nurse game. Am J Nurs. 1968;68(1):101–5.
7. Stein LI, Watts DT, Howell T. The doctor-nurse game revisited. N Engl J Med. 1990;322(8):546–9. https://doi.org/10.1056/NEJM199002223220810.
8. Davis AJ, Fowler M, Aroskar M. Ethical dilemmas & nursing. 5th ed. Boston: Pearson; 2010.
9. Kramer M. Reality shock: why nurses leave nursing. St. Louis: C.V. Mosby; 1974.
10. Vaughn RH. The actual incidence of moral problems in nursing: a preliminary study in empirical ethics. Washington DC: Catholic University of America; 1935.
11. Lemkuhl A. Theologia Moralis (2 vols. cpl/2 Bande)- Vol 1. Theologiam moralem generalem et ex speciali theologia morali tractatus de virtutibus et officiis vitae christianae/Vol, II. Theologiae moralis specialis parten secundam seu tractatus De subsidiis vitae christianae cum duabus appendicibus. Herder: Frigurgi Brisgoviae; 1910.
12. Parsons SE. Nursing problems and obligations. Boston: M. Barrows & Co.; 1928. p. 109.
13. American Nurses' Association (ANA). Code of ethics for nurses with interpretive statements. Silver Spring: American Nurses Association; 2015. p. 37.
14. Robb IAH. Nursing ethics: for hospital and private use. New York: Koeckert; 1900. p. 256–7.
15. Hirschman AO. Exit, voice, and loyalty: responses to decline in firms, organizations and states. Cambridge: Harvard University Press; 1970.
16. Donley R, Flaherty MJ. Revisiting the American Nurses Association's first position on education for nurses. Online J Issues Nurs. 2008;13(2). http://www.nursingworld.org/MainMenuCategories/ANAMarketplace/ANAPeriodicals/OJIN/TableofContents/vol132008/No2May08/ArticlePreviousTopic/EntryIntoPracticeUpdate.html
17. American Nurses' Association (ANA). Education for nursing. Am J Nurs. 1965;65(12):106–11.
18. Fowler MD. Heritage ethics: toward a thicker account of nursing ethics. Nurs Ethics. 2016;23(1):7–21. Published online before print November 23, 2015
19. Fowler M. Why the history of nursing ethics matters. Nurs Ethics. 2017;24(3):292–304.
20. Fowler M. Guide to nursing's social policy statement: understanding the essence of the profession from social contract to social covenant. Silver Spring: American Nurses Association; 2015.
21. Sullivan WM. Work and integrity: the crisis and promise of professionalism in America. 2nd ed. San Francisco, CA: Jossey-Bass; 2005. p. 2–3.
22. Sullivan W, Benner P. Challenges to professionalism: work integrity and the call to renew and strengthen the social contract of the professions. Am J Crit Care. 2005;14(1):79.
23. Johnstone MJ, Hutchinson A. 'Moral distress' – time to abandon a flawed nursing construct? Nurs Ethics. 2015;22(1):10.
24. Fowler M, Mahon K. Moral development and clinical decision making. Nurs Clin North Am. 1979;14(1):3–12. Publication of authorship correction in subsequent issue
25. McCarthy J, Gastmans C. Moral distress: a review of the argument-based nursing ethics literature. Nurs Ethics. 2015;22(1):131–52. https://doi.org/10.1177/0969733014557139.
26. McCarthy J, Deady R. Moral distress reconsidered. Nurs Ethics. 2008;15(2):254–62.
27. Peter E, Liaschenko J. Moral distress reexamined: a feminist interpretation of nurses' identities, relationships, and responsibilites. J Bioeth Inq. 2013;10(3):337–45. https://doi.org/10.1007/s11673-013-9456-5.
28. Peter E. Guest editorial: three recommendations for the future of moral distress scholarship. Nurs Ethics. 2015;22(1):3–4.
29. Jameton A. What moral distress in nursing history could suggest about the future of health care. AMA J Ethics. 2017;19(6):617–28.
30. Baer ED, D'Antonio P, Rinker S, Lanaugh JE. Enduring issues in American nursing. New York: Springer; 2002. p. 337.
31. Craig SL, Muskat B. Bouncers, brokers, and glue: the self-described roles of social workers in urban hospitals. Health Soc Work. 2013;38(1):7–16.
32. Gregorian C. A career in hospital social work: do you have what it takes? Soc Work Health Care. 2005;40(3):1–14. https://doi.org/10.1300/J010v40n03_01.
33. Congress E. What social workers should know about ethics: understanding and resolving practice dilemmas. Adv Soc Work. 2000;1(1):1–22.

34. Mattison M. Ethical decision making: the person in the process. Soc Work. 2000;45(3):201–12.
35. Weinberg M. Moral distress: a missing but relevant concept for ethics in social work. Can Soc Work Rev. 2009;26(2):139–51.
36. Epstein E, Delgado S. Understanding and addressing moral distress. Online J Issues Nurs. 2010;15(3):1. Manuscript 1.
37. Fourie C. Moral distress and moral conflict in clinical ethics. Bioethics. 2015;29(2):91–7. https://doi.org/10.1111/bioe.12064.
38. Hamric AB. A case study of moral distress. J Hosp Palliat Care. 2014;16(8):457–63.
39. Hamric AB, Blackhall LJ. Nurse-physician perspectives on the care of dying patients in intensive care units: collaboration, moral distress, and ethical climate. Crit Care Med. 2007;35(2):422–9. https://doi.org/10.1097/01.CCM.0000254722.50608.2D.
40. Hamric AB, Borchers CT, Epstein EG. Development and testing of an instrument to measure moral distress in healthcare professionals. AJOB Prim Res. 2012;3(2):1–9.
41. Varcoe C, Pauly B, Storch J, Newton L, Makaroff K. Nurses' perceptions of and responses to morally distressing situations. Nurs Ethics. 2012;19(4):488–500. https://doi.org/10.1177/0969733011436025.
42. Wiegand DL, Funk M. Consequences of clinical situations that cause critical care nurses to experience moral distress. Nurs Ethics. 2012;19(4):479–87. https://doi.org/10.1177/0969733011429342.
43. Allen R, Judkins-Cohn T, deVelasco R, Forges E, Lee R, Clark L, Procunier M. Moral distress among healthcare professionals at a health system. JONAS Healthc Law Ethics Regul. 2013;15(3):111–118; . quiz 119-120. https://doi.org/10.1097/NHL.0b013e3182a1bf33.
44. de Veer AJ, Francke AL, Struijs A, Willems DL. Determinants of moral distress in daily nursing practice: a cross sectional correlational questionnaire survey. Int J Nurs Stud. 2013;50(1):100–8. https://doi.org/10.1016/j.ijnurstu.2012.08.017.
45. Moffatt M. Reducing moral distress in case managers. Prof Case Manag. 2014;19(4):173–86.
46. Trotochaud K, Coleman JR, Krawiecki N, McCracken C. Moral distress in pediatric healthcare providers. J Pediatr Nurs. 2015;30(6):908–14. https://doi.org/10.1016/j.pedn.2015.03.001.
47. Bruce CR, Miller SM, Zimmerman JL. A qualitative study exploring moral distress in the ICU team: the importance of unit functionality and intrateam dynamics. Crit Care Med. 2015;43(4):823–31. https://doi.org/10.1097/CCM.0000000000000822.
48. Houston S, Casanova MA, Leveille M, Schmidt KL, Barnes SA, Trungale KR, Fine RL. The intensity and frequency of moral distress among different healthcare disciplines. J Clin Ethics. 2013;24(2):98–112. https://www.ninr.nih.gov.
49. Manttari-van der Kuip M. Moral distress among social workers: the role of insufficient resources. Int J Soc Welf. 2016;25(1):86–97.
50. Ulrich C, O'Donnell P, Taylor C, Farrar A, Danis M, Grady C. Ethical climate, ethics stress, and the job satisfaction of nurses and social workers in the United States. Soc Sci Med. 2007;65(8):1708–19. https://doi.org/10.1016/j.socscimed.2007.05.050.
51. Blomberg H, Kallio J, Kroll C, Saarinen A. Job stress among social workers: determinants and attitude effects in the Nordic countries. Br J Soc Work. 2015;45(7):2089–105.
52. Huxley P, Evans S, Gately C, Webber M, Mears A, Pajak S, Kendall T, Medina J, Katona C. Stress and pressures in mental health social work: the worker speaks. Br J Soc Work. 2005;35(7):1063–79.
53. Kim H, Stoner M. Burnout and turnover intention among social workers: effects of role stress, job autonomy and social support. Adm Soc Work. 2008;32(3):5–25.
54. Rossi A, Cetrano G, Pertile R, Rabbi L, Donisi V, Grigoletti L, Curtolo C, Tansella M, Thornicroft G, Amaddeo F. Burnout, compassion fatigue, and compassion satisfaction among staff in community-based mental health services. Psychiatry Res. 2012;200(2–3):933–8.
55. DiFranks NN. Social workers and the NASW code of ethics: belief, behavior, disjuncture. Soc Work. 2008;53(2):167–76.
56. Fenton J. An analysis of 'ethical stress' in criminal justice social work in Scotland: the place of values. Br J Soc Work. 2015;45:1415–32.

57. Taylor MF. Professional dissonance: a promising concept for clinical social work. Smith College Studies in Social Work. 2007;77(1):89–99.
58. Lev S, Ayalon L. Moral distress among long-term care social workers: questionnaire validation. Research on social work practice from Online First. 2016. Retrieved from http://journals.sagepub.com/doi/abs/10.1177/1049731516672070.
59. Corley MC, Elswick RK, Gorman M, Clor T. Development and evaluation of a moral distress scale. J Adv Nurs. 2001;33(2):250–6.
60. Hamric AB. Empirical research on moral distress: issues, challenges, and opportunities. HEC Forum. 2012;24(1):39–49. https://doi.org/10.1007/s10730-012-9177-x.
61. Fantus S, Greenberg RA, Muskat B, Katz D. Exploring moral distress for hospital social workers. The British Journal of Social Work. 2017. Online First https://doi.org/10/1093/bjsw/bcw113.
62. Cullen AF. Leaders in our own lives: suggested indications for social work leadership from a study of social work practice in a palliative care setting. Br J Soc Work. 2013;43(8):1527–44.
63. Otis-Green S, Sidhu RK, Del Ferraro C, Ferrell B. Integrating social work into palliative care for lung cancer patients and families: a multidimensional approach. J Psychosoc Oncol. 2014;32(4):431–46. https://doi.org/10.1080/07347332.2014.917140.
64. Stein GL, Fineberg IC. Advance care planning in the USA and UK: a comparative analysis of policy, implementation and the social work role. Br J Soc Work. 2013;43(2):233–48.
65. Kalliath P, Kalliath T. Does job satisfaction mediate the relationship between work-family conflict and psychological strain? A study of Australian social workers. Asia Pac J Soc Work Dev. 2013;23(2):91–105.
66. Manttari-van der Kuip M. The deteriorating work-related well-being among statutory social workers in a rigorous economic context. Eur J Soc Work. 2014;17(5):672–88.
67. Sutinen R, Kivimaki M, Elovainio M, Virtanen M. Organizational fairness and psychological distress in hospital physicians. Scand J Public Health. 2002;30(3):209–15. https://doi.org/10.1080/14034940210133843.
68. Auerbach C, Mason SE, Heft Laporte H. Evidence that supports the value of social work in hospitals. Soc Work Health Care. 2007;44(4):17–32. https://doi.org/10.1300/J010v44n04_02.
69. Lloyd C, King R, Chenoweth L. Social work, stress and burnout: a review. J Ment Health. 2002;11(3):255–65.
70. Gibson M. Social worker shame: a scoping review. Br J Soc Work. 2016;46(2):549–65.
71. Jansson BS, Dodd SJ. Ethical activism: strategies for empowering medical social workers. Soc Work Health Care. 2002;36(1):11–28. https://doi.org/10.1300/J010v36n01_02.
72. Beddoe L, Davys AM, Adamson C. Never trust anybody who says "I don't need supervision": practitioners' beliefs about social worker resilience. Pract Soc Work Action. 2014;26(2):113–30.
73. Bogo M, Paterson J, Tufford L, King R. Supporting front-line practitioners' professional development and job satisfaction in mental health and addiction. J Interprof Care. 2011;25(3):209–14. https://doi.org/10.3109/13561820.2011.554240.
74. Landau R. Ethical dilemmas in general hospitals: social workers' contribution to ethical decision-making. Soc Work Health Care. 2000;32(2):75–92.
75. Acker GM. The challenges in providing services to clients with mental illness: managed care, burnout and somatic symptoms among social workers. Community Ment Health J. 2010;46(6):591–600. https://doi.org/10.1007/s10597-009-9269-5.
76. Leichtentritt RD. Beyond favourable attitudes of end-of-life rights: the experiences of Israeli health care social workers. Br J Soc Work. 2011;41(8):1459–76.
77. Sabin JE. Using moral distress for organizational improvement. J Clin Ethics. 2017;28(1):33–6.
78. Whitehead PB, Herbertson RK, Hamric AB, Epstein EG, Fisher JM. Moral distress among healthcare professionals: report of an institution-wide survey. J Nurs Scholarsh Off Publ Sigma Theta Tau Int Honor Soc Nurs. 2015;47(2):117–25. https://doi.org/10.1111/jnu.12115.
79. Shanafelt TD, Balch CM, Bechamps G, et al. Burnout and medical errors among American surgeons. Ann Surg. 2010;251(6):995–1000. https://doi.org/10.1097/SLA.0b013e3181bfdab3.

80. Fuks A, Brawer J, Boudreau JD. The foundation of physicianship. Perspect Biol Med. 2012;55(1):114–26. https://doi.org/10.1353/pbm.2012.0002.

81. Minogue B. The two fundamental duties of the physician. Acad Med J Assoc Am Med Coll. 2000;75(5):431–42.

82. Cruess SR, Cruess RL. Professionalism and medicine's social contract with society. Virtual Mentor VM. 2004;6(4). https://doi.org/10.1001/virtualmentor.2004.6.4.msoc1-0404.

83. Long C, Tsay EL, Jacobo SA, Popat R, Singh K, Chang RT. Factors associated with patient press ganey satisfaction scores for ophthalmology patients. Ophthalmology. 2016;123(2):242–7. https://doi.org/10.1016/j.ophtha.2015.09.044.

84. Sinsky C, Colligan L, Li L, et al. Allocation of physician time in ambulatory practice: a time and motion study in 4 specialties. Ann Intern Med. 2016;165(11):753–60. https://doi.org/10.7326/M16-0961.

85. Rushton CH, Kaszniak AW, Halifax JS. A framework for understanding moral distress among palliative care clinicians. J Palliat Med. 2013;16(9):1074–9. https://doi.org/10.1089/jpm.2012.0490.

86. Austin CL, Saylor R, Finley PJ. Moral distress in physicians and nurses: impact on professional quality of life and turnover. Psychol Trauma Theory Res Pract Policy. 2017;9(4):399–406. https://doi.org/10.1037/tra0000201.

87. Fumis RRL, Junqueira Amarante GA, de Fátima Nascimento A, Vieira Junior JM. Moral distress and its contribution to the development of burnout syndrome among critical care providers. Ann Intensive Care. 2017;7(1):71. https://doi.org/10.1186/s13613-017-0293-2.

88. Wurm W, Vogel K, Holl A, et al. Depression-burnout overlap in physicians. PLoS One. 2016;11(3):e0149913. https://doi.org/10.1371/journal.pone.0149913. van Wouwe J, ed.

89. Bianchi R, Schonfeld IS, Laurent E. Physician burnout is better conceptualised as depression. Lancet Lond Engl. 2017;389(10077):1397–8. https://doi.org/10.1016/S0140-6736(17)30897-8.

90. Center C, Davis M, Detre T, et al. Confronting depression and suicide in physicians: a consensus statement. JAMA. 2003;289(23):3161–6. https://doi.org/10.1001/jama.289.23.3161.

91. Campbell SM, Ulrich CM, Grady C. A broader understanding of moral distress. Am J Bioeth AJOB. 2016;16(12):2–9. https://doi.org/10.1080/15265161.2016.1239782.

92. Larson CP, Dryden-Palmer KD, Gibbons C, Parshuram CS. Moral distress in PICU and neonatal ICU practitioners: a cross-sectional evaluation. Pediatr Crit Care Med. 2017;18(8):e318–26. https://doi.org/10.1097/PCC.0000000000001219.

93. Garros D, Austin W, Carnevale FA. Moral distress in pediatric intensive care. JAMA Pediatr. 2015;169(10):885–6. https://doi.org/10.1001/jamapediatrics.2015.1663.

94. Brandon D, Ryan D, Sloane R, Docherty SL. Impact of a pediatric quality of life program on providers' moral distress. MCN Am J Matern Child Nurs. 2014;39(3):189–97.

95. Rosenberg AR, Orellana L, Kang TI, Geyer JR, Feudtner C, Dussel V, Wolfe J. Differences in parent-provider concordance regarding prognosis and goals of care among children with advanced cancer. J Clin Oncol. 2014;32(27):3005–11. https://doi.org/10.1200/JCO.2014.55.4659.

96. Hilden JM, Emanuel EJ, Fairclough DL, Link MP, Foley KM, Clarridge BC, et al. Attitudes and practices among pediatric oncologists regarding end-of-life care: results of the 1998 American Society of Clinical Oncology survey. J Clin Oncol. 2001;19(1):205–12. https://doi.org/10.1200/JCO.2001.19.1.205.

97. Mack JW, Smith TJ. Reasons why physicians do not have discussions about poor prognosis, why it matters, and what can be improved. J Clin Oncol. 2012;30(22):2715–7. https://doi.org/10.1200/JCO.2012.42.4564.

98. Mack JW, Wolfe J, Cook EF, Grier HE, Cleary PD, Weeks JC. Hope and prognostic disclosure. J Clin Oncol. 2007;25(35):5636–42. https://doi.org/10.1200/JCO.2007.12.6110.

99. Kaye E, Mack JW. Parent perceptions of the quality of information received about a child's cancer. Pediatr Blood Cancer. 2013;60(11):1896–901. https://doi.org/10.1002/pbc.24652.

100. Mack JW, Wolfe J, Grier HE, Cleary PD, Weeks JC. Communication about prognosis between parents and physicians of children with cancer: parent preferences and the impact

of prognostic information. J Clin Oncol. 2006;24(33):5265–70. https://doi.org/10.1200/JCO.2006.06.5326.

101. Whitehead K. Motherhood as a gendered entitlement: intentionality, "othering," and homosociality in the online infertility community. Can Rev Sociol. 2016;53(1):94–122. https://doi.org/10.1111/cars.12093.

102. Hill DL, Miller V, Walter JK, Carroll KW, Morrison WE, Munson DA, et al. Regoaling: a conceptual model of how parents of children with serious illness change medical care goals. BMC Palliat Care. 2014;13(1):9. https://doi.org/10.1186/1472-684X-13-9.

103. Bronfenbrenner U. In: Bronfenbrenner U, editor. Making human beings human: bioecological perspectives on human development. Thousand Oaks: Sage; 2005.

104. Kazak AE. Comprehensive care for children with cancer and their families: a social ecological framework guiding research, practice, and policy. Child Serv Soc Policy Res Pract. 2001;4(4):217–33.

105. Mooney-Doyle K, Deatrick JA. Parenting in the face of childhood life-threatening conditions: the ordinary in the context of the extraordinary. Palliat Support Care. 2016;14(3):187–98. https://doi.org/10.1017/S1478951515000905.

106. Feudtner C, Walter JK, Faerber JA, Hill DL, Carroll KW, Mollen CJ, et al. Good-parent beliefs of parents of seriously ill children. JAMA Pediatr. 2015;169(1):39–47. https://doi.org/10.1001/jamapediatrics.2014.2341.

107. Hinds PS, Oakes LL, Hicks J, Powell B, Srivastava DK, Spunt SL, et al. "Trying to be a good parent" as defined by interviews with parents who made phase I, terminal care, and resuscitation decisions for their children. J Clin Oncol. 2009;27(35):5979–85. https://doi.org/10.1200/JCO.2008.20.0204.

108. Mooney-Doyle K, dos Santos MR, Szylit R, Deatrick JA. Parental expectations of support from healthcare providers during pediatric life-threatening illness: a secondary, qualitative analysis. J Pediatr Nurs. 2017;36:163–72.

109. Laing CM, Moules NJ, Estefan A, Lang M. Stories that heal: understanding the effects of creating digital stories with pediatric and adolescent/young adult oncology patients. J Pediatr Oncol Nurs. 2017;34(4):272–82. https://doi.org/10.1177/1043454216688639.

Stephen M. Campbell, Connie M. Ulrich,
and Christine Grady

Moral distress has become a well-established issue of concern in the nursing literature and is increasingly getting attention in other domains of healthcare.[1] According to Andrew Jameton, who first introduced the topic in the 1980s, "*Moral distress* arises when one knows the right thing to do, but institutional constraints make it nearly impossible to pursue the right course of action" [1, p. 6]. Since the time of this initial characterization, the phenomenon of moral distress has been discussed, defined, and researched by several authors. While there are subtle variations in how different authors have understood it, the following are widely held to be defining elements of moral distress[2]:

1. It arises when one believes one knows the morally right thing to do (or avoid doing), but one's ability to do this is constrained by internal and/or external factors.

---

[1] "A Broader Understanding of Moral Distress" was republished with the permission of the American Journal of Bioethics 2016 Dec;16(12):2-9. Taylor & Francis Group, LLC.

[2] For some representative characterizations of moral distress that include one or more of these features, see Jameton [1], Wilkinson [2], Jameton [3], Webster and Baylis [4, p. 218], Corley [5], Hanna [6], Rushton [7], American Association of Critical-Care Nurses [8], Canadian Nurses Association [9], McCarthy and Deady [10], Epstein and Hamric [11], Austin et al. [12], Chen [13], Ulrich et al. [14], Epstein and Delgado [15], and Hamric [16].

S.M. Campbell (✉)
Department of Philosophy, Bentley University, Waltham, MA, USA
e-mail: S.Campbell80@gmail.com

C.M. Ulrich
Lillian S. Brunner Endowed Chair, University of Pennsylvania School of Nursing,
Philadelphia, PA, USA

Department of Medical Ethics and Health Policy, Perelman School of Medicine,
University of Pennsylvania School of Medicine, Philadelphia, PA, USA

C. Grady
Department of Bioethics, National Institutes of Health, Bethesda, MD, USA

© Springer International Publishing AG 2018
C.M. Ulrich, C. Grady (eds.), *Moral Distress in the Health Professions*,
https://doi.org/10.1007/978-3-319-64626-8_4

2. It comes in two phases. There is "initial distress" at the time of potential action (or inaction); later, there is "reactive distress" or "moral residue" that occurs in response to the initial episode of moral distress.
3. It involves the compromising of one's moral integrity or the violation of one's core values.

This is the prevailing understanding of what moral distress is.

Our purpose in this essay is to motivate a broader understanding of moral distress. There is a wider range of cases that can sensibly be framed as moral distress, and it is important to recognize them as such. Embracing a broader conception of moral distress does not in any way undermine the relevance or importance of the groundbreaking work that has been done on this topic over the past several years. It simply implies that this previous work has focused on one type of moral distress. In the first section, we present six cases that fall outside bounds of the traditional characterization of moral distress. We argue that it is desirable for a definition of moral distress to encompass them. In the second section, we propose a new definition that accommodates all six cases, as well as the cases accommodated by the traditional definition of moral distress. In the third section, we respond to worries that this new definition is overly broad. In the fourth section, we take some first steps toward the development of a taxonomy of moral distress.

4.1   The Case for Broadening Our Understanding of Moral Distress

The purpose of this section is to motivate the need to broaden the traditional characterization of moral distress. Our strategy is to present six cases of distress and explain why they should be understood to be forms of moral distress. It should first be clarified that the inclusion of our six cases cannot be motivated by an appeal to the meaning of "moral distress." Although the words "moral" and "distress" are pieces of natural language, the phrase "moral distress" is a term of art. It was first coined in 1984 for the purpose of naming a phenomenon that was observed in nursing practice, and the phrase has had life almost exclusively within the medical and bioethics literature. For this reason, we are happy to grant that "moral distress" means whatever the scholars writing about it have taken it to mean. Since the prevailing understanding of moral distress in the relevant literature excludes the six cases, a brute appeal to meaning does nothing to motivate their inclusion.

What can serve to motivate the inclusion of the six cases is reflection on the features of moral distress that help to explain why health care professionals and bioethicists have had a sustained interest in this topic. The phenomenon of moral distress is, first and foremost, a practical problem. In the nursing profession, its problematic nature largely consists in its adverse effects on the well-being of nurses, the quality of patient care, and nurse retention [5, 17–21]. Of course, it is likely that any type of on-the-job distress or frustration can contribute to such problems—including, for instance, distress that stems from having an overbearing co-worker or being continually exposed to the suffering and death of patients. But there is something especially problematic and worrisome about distress that arises when individuals feel morally compromised or tainted in some way. As we see it, this is what distinguishes moral distress from other kinds of distress. This may explain why the topic of moral distress, as opposed to mere distress, has received so much attention. Arguably, bioethicists, policymakers, and health care administrators have special reason to try to eliminate, or at least mitigate, this kind of distress in health care contexts. Admittedly, this talk of being "morally compromised" or "morally tainted" is rather vague. Even so, these expressions are useful starting points for thinking about what moral distress is and why it is often important to address it.

Although none of our six cases involves moral distress as traditionally conceived, each involves an individual in a health care context experiencing distress because he or she feels morally compromised in some way. Each of the cases describes a type of experience that certainly could, and probably often does, contribute to a loss of well-being, a diminishment of job satisfaction, poorer job performance, and burnout. As we hope to show, there appears to be no principled reason why a definition of moral distress should exclude these cases.

4.1.1   Moral Uncertainty

A newly appointed general surgeon who has just finished his residency training is assigned to a disproportionate share of the Medicaid and uninsured cases. These patients are complicated, and many of them suffer with multiple comorbidities due to limited access to primary care and treatment. In fact, several of his patients have already experienced postoperative complications following gastrointestinal surgery, including abdominal sepsis and evisceration. He feels that assigning a new surgeon to these patients is unfair to them since he has less experience than other surgeons. He worries that he might be harming them. He is distressed about this but does not know the best way to respond. One option is to simply continue doing the surgeries to the best of his ability. Alternatively, he could complain to his superiors, though he is worried that he might be labeled as a troublemaker and that some of the surgeries might get delayed. He also considers seeking the advice of a more senior surgeon, but he suspects that he would be told that this is the way the system works and it is good training. He is not sure what the morally best course of action is.

Moral distress is commonly thought to arise only in cases where a person thinks she knows the morally right course of action. No doubt, there are times when we find ourselves in situations in which we think we know exactly what morality demands of us. Still, as illustrated in the case just described, it is all too common that we fall short of having such knowledge. Life as a moral agent is complex, and it is often difficult or impossible to know what the morally right course of action is. One reason for this is that it is no easy matter saying which moral theory or moral principles are correct. Even moral philosophers, whose careers revolve around thinking about ethics, are continually developing, revising, and fine-tuning their own views about morality. Another reason has to do with uncertainty or indeterminacy concerning the professional duties or proper role of different health care professionals and workers. A final reason for moral uncertainty is that we often lack pertinent empirical information about our situation. To appreciate this point, imagine a situation in which a patient is about to consent to a procedure without having adequate information. Jameton [3] gives the following (nonexhaustive) list of ways in which a nurse might immediately react to this potentially distressing situation:

- Just relate the information to the patient.
- Ask the physician leading questions to elicit the information.
- Step aside with the physician and suggest that he or she reconsider the procedure, or suggest that the physician or nurse give the patient more information.
- Call in the head nurse.
- Resign on the spot.
- Scream.
- Undermine the process.
- Say a prayer.
- Do nothing.

Jameton goes on to list a host of other possible actions the nurse might take soon after the event or that he or she might take if this sort of situation arises regularly (pp. 544–45). Given the vast array of possible actions open to us at any given moment, it is no wonder that we often fall short of knowing what the morally right action would be. We often will not know all of the possible actions that are available to us, much less what consequences they would all have—and, as a result, will not know what the right thing to do is. Yet, even in the absence of such knowledge, it is possible for one to experience negative attitudes like guilt or unease. One might have a firm conviction that one did the *wrong* thing without having the faintest clue what the right action would have been. Or one might simply *suspect* that one failed to do the morally right thing, even if one is not at all sure. Distress in the form of guilt or self-criticism can arise under such circumstances.

4.1.2   Mild Distress

An operating-room scrub nurse is frequently assigned to work with a pediatric cardiac surgeon who has a reputation for explosive outbursts in the operating room (OR). The surgeon has screamed profanities at the heart-bypass perfusion team, anesthesiologists, residents, and nurses, and has even been known to throw instruments across the room. The scrub nurse happens to be in the surgeon's good graces and is one of the few people immune to her outbursts. Even so, he finds it troubling to see his colleagues berated and thinks he should intervene in some way. Yet, when these outbursts happen, he feels constrained from saying anything to the surgeon for fear of falling out of favor with her and possibly making the situation worse, which might undermine the cooperation and teamwork needed to save the health or life of the child on the operating-room table. Taken in isolation, each episode is only mildly upsetting to the scrub nurse. Indeed, the first time he experienced the surgeon's behavior, he just rolled his eyes and continued to focus on his work. But these instances of distress have a negative cumulative effect over time.

Discussions of moral distress often give the impression that every episode of moral distress is a dramatic, life-altering affair. The common practice of associating moral distress with the compromise of one's moral integrity, the violation of one's core values, or even the threatening of one's very identity suggests serious moral compromise. And the word "distress" could easily be taken to denote a very strong emotional reaction.

We grant that the most disturbing and significant instances of moral distress will be those that create intense feelings of distress and shake people to their "moral core" by violating their core values or compromising their moral integrity. However, individuals can be morally compromised in less momentous ways that are still damaging and worth addressing (cf. [5], p. 637). It can be distressing to be prevented from doing what you think is the morally right thing to do even when the action in question has nothing to do with your core values. People who are morally corrupt can have rare moments of moral conscientiousness and can experience distress if they are kept from acting rightly—even if they do not really have any moral integrity to compromise. Finally, as exemplified in the case just described, there are occasions on which a person finds it only mildly distressing that she is constrained from doing what she thinks is morally best. Episodes of mild distress, when they occur on a regular basis, can have an adverse cumulative effect on those who experience them. The difference between strong and mild distress is a difference only in degree, not in kind. Rather than denying the existence of mild moral distress, we should simply recognize that isolated instances of mild moral distress have lower moral priority than stronger forms.

4.1.3   Delayed Distress

An experienced emergency physician is on duty when a 55-year-old female patient arrives at the emergency department via ambulance after being ejected from the vehicle during a roll over motor vehicle collision. She sustained multiple fractures, severe facial injuries, and a significant closed head injury. The patient was intubated on scene by the paramedics, and cardiopulmonary resuscitation (CPR) is in progress after a traumatic cardiac arrest. Upon arrival, she has no pupillary response and she remains in full cardiac arrest. The emergency physician, nurses, and trauma team immediately continue resuscitation, placing multiple lines, giving meds and blood products in accordance with Advanced Trauma Life Support protocols. Multiple units of blood, IV fluids, and medications are administered in an attempt to get return of spontaneous circulation and manage the patient's injury. After 30 minutes of aggressive resuscitation, the patient has return of spontaneous circulation and is transferred to the OR for a craniotomy to relieve intracranial pressure, which helps stabilize her condition although she is still in critical condition and the team questions the likelihood of a meaningful recovery. On his drive home, the physician begins to reflect on the attempts to resuscitate the woman and is troubled that they were so aggressive for so long. His knowledge and experience told him the chances of a meaningful recovery from her devastating injuries were very low. He wonders about the woman's quality of life and whether aggressive resuscitation was the best option.

On the traditional picture of moral distress, it comes in two stages. First, there is *initial distress*, which is felt at the very time at which one's action is constrained by internal or external factors. This is followed by *reactive distress*, or what some have called "moral residue." It appears to be widely assumed that both initial distress and reactive distress are essential elements of moral distress. Indeed, Jameton, one of the first to explicitly draw the initial/reactive distress distinction, defines reactive distress in terms of initial distress: "Reactive distress is the distress that people feel when they do not act upon their initial distress" [3, p. 544]. This characterization of reactive distress presupposes the existence of initial distress.[3]

As our case of delayed distress illustrates, it is perfectly possible for a person to fail to have distress at the time of being morally comprised. In emergency situations, the urgent need for action can prevent a person from fully processing the nature of the situation and her actions and, as a result, from feeling the appropriate emotions. Yet, if later reflection leads a person to recognize that she had been constrained from acting in the morally best way and if she feels distress as a result of this, there is no reason why we should not treat this as a case of moral distress. Such a person might be in a mental state nearly identical to that of a morally distressed person who did have initial distress.

We can also imagine scenarios in which one experiences initial distress without reactive distress. After the period of initial distress, any number of events might prevent an individual from experiencing reactive distress. One might forget about the distressing episode (particularly in cases of mild distress), repress the memory of it, formulate a post hoc rationalization of one's behavior, or become occupied with more pressing concerns (such as the death of a loved one). An individual might not have reactive distress because she comes to see her initial distress as inappropriate—perhaps because she gains more information that leads her to revise her moral assessment of the situation. There are also cases where the experience of reactive distress is precluded by a medical condition or death.

In light of these considerations, it is a mistake to insist that initial distress and moral residue are necessary features of moral distress. Cases in which only one or the other occurs still deserve to be treated as cases of moral distress.

4.1.4   Moral Dilemma

A bioethics consultant is called by the pediatric oncology team to get advice about a 13-year-old patient with cancer whose clinical situation is precarious. The team members want to know what they should tell the patient about his diagnosis and prognosis. When the patient was diagnosed over a year ago, his parents were worried that knowledge of his condition would be overwhelming and cause him unnecessary distress. They asked the team not to give him details about his disease. Despite many months of aggressive treatment, his cancer is progressing and he is experiencing some debilitating complications from the treatment. The parents are still adamant that he should not be given details about his condition, and the team does not know how to respond. After meeting with the patient and his parents, the bioethics consultant feels torn between respecting the wishes of the parents who know their son and have his best interests at heart, and showing respect for the patient and his welfare by advising the team to disclose what they know about his situation that might help him make informed decisions. Each option seems morally regrettable: Either they deceive this patient about his condition, or they violate the parents' wishes and give the patient information that is likely to cause him distress. The bioethics consultant thinks there are

---

[3] See also [4, p. 218].

equally good arguments to be made against each of these choices. He ultimately recommends disclosing information to the patient, but he feels some guilt about making this recommendation.

When Jameton first introduced the phenomenon of moral distress, he contrasted it with two other kinds of cases: cases of moral uncertainty, and moral dilemmas. We have already challenged the idea that moral distress and moral uncertainty are mutually exclusive phenomena. We now wish to suggest that there is also some overlap between moral dilemmas and moral distress. Moral dilemmas "arise when two (or more) clear moral principles apply, but they support mutually inconsistent courses of action" [1, p. 6]. Thus, if moral distress (as traditionally conceived) arises in cases where morality pulls a person in one direction but constraints pull her in another, moral dilemmas are cases in which morality itself pulls a person in competing directions. As a result, dilemmas are cases in which one cannot avoid doing something morally regrettable.

It seems a mistake to define moral distress in such a way that it cannot be experienced in moral dilemmas. Moral dilemmas are classic cases in which people do, and arguably should, feel morally compromised. Distress is a natural response to a situation in which you are "damned if you do, damned if you don't."[4] To make space for the possibility of moral distress in response to a moral dilemma, we should simply reject the idea that moral distress only results when one is constrained from doing the morally right thing. Moral dilemmas are situations in which there is no (purely) morally right thing to do. Being thrust into a moral dilemma can lead to feelings of distress and moral compromise, loss of well-being, and so on, just as naturally as being faced with a morally right option that one is kept from taking.

4.1.5 Bad Moral Luck

A psychiatrist pushes hard to get his patient to take a medication that he believes will help to address her intractable depression. The patient is initially reluctant, but he eventually persuades her. Two weeks later, she takes an intentional overdose of the medication, which results in her death. The psychiatrist feels terrible about his role in the patient's suicide and wonders whether he did the right thing. However, after reviewing the case, he continues to think that he did exactly what someone in his position should have done, given the evidence available to him at the time. Even so, he feels great distress about the consequences of his action.

One of the most firmly established beliefs about moral distress is that it is always the result of an individual failing to do the morally right thing where this failing is the result of internal and external constraints. However, it is possible for one to feel morally compromised or tainted even in cases where one is not constrained and one successfully performs what one judges to be the morally best action. One type of

---

[4] For an influential discussion of moral dilemmas and the appropriateness of one species of distress ("agent-regret"), see Williams [22].

case that fits this description involves a certain species of "moral luck."[5] As illustrated in the case just described, sometimes individuals perform what they deem to be the morally best action based on the best information and evidence available to them at the time, without any internal or external constraints. Yet their actions, in conjunction with factors beyond their control, turn out to have morally undesirable consequences (such as the suffering or death of another sentient being or the violation of someone's autonomy or rights). This can lead to feelings of distress. One need not think she should have acted differently. The person may firmly believe that it would have been wrong of her to do otherwise, given what she knew at the time. Still, she may feel terrible that she played a role in bringing about a morally regrettable outcome. This is an instance of distress rooted in the sense that one has been morally compromised, despite the fact that one is not guilty of acting in a wrong or blameworthy manner.

4.1.6   Distress by Association

> A nurse at the bedside is responsible for providing clinical care to her patient, who is also a participant in a research study. Based on conversations with the patient, she feels that the patient does not really understand the purpose of the research study and is desperately hoping for any benefit to extend his life. The patient tells her that he did not read the consent form carefully. As the study progresses, the patient's clinical status begins to deteriorate, yet he wants to continue on the study because he thinks it will benefit him. The nurse believes that the patient's continued participation in this research study is morally wrong. She encourages him to meet with the research team to discuss his clinical situation and the purpose and progress of the research, but he is uninterested in doing that. When she mentions her concerns to the research team, they respond that he understands the study well enough and that she should stop worrying. The nurse becomes increasingly troubled by her interactions with this patient. Although she is not part of the research team, she has responsibilities for caring for him and monitoring his clinical status at the bedside where research procedures occur. She feels guilty and distressed about her involvement despite the fact that she has tried very hard to remedy the situation.

This is a second type of case in which one is not subjected to internal or external constraints and one does not fail to do the morally right thing. Distress by association is not grounded in one's own action or omission but in one's association with another party—which might be one or more individuals, or a collective entity.[6] As we are understanding it, distress by association is not essentially a matter of emotional contagion, where distress in one person is triggered by exposure to another's distress. Nor is it a matter of empathetically experiencing the distress that someone else is, or should be, experiencing. Instead, distress by association springs from the sense of being morally compromised due to one's connection with some other party. Perhaps this other party acted immorally with malicious intentions, or acted negli-

---

[5] The *locus classicus* for this topic is Williams and Nagel [23]. Our present focus is on what is often called "resultant luck," or luck in how things turn out.

[6] This idea is sometimes explored in discussions of "moral taint." See, for instance, Oshana [24].

gently, or acted morally but with morally disastrous results. Or perhaps this other party has morally condemnable beliefs, attitudes, or motives, without being guilty of *acting* in morally questionable ways. In some cases, distress by association concerns one's membership in a group or organization that has caused a morally undesirable state of affairs, though the responsibility for this does not fall on the distressed individual—or, perhaps, on any particular individual. A doctor might experience distress by association because she works in a health care facility that does not provide adequate care or quality of life for its patients. Here, as in all of the previous cases, it makes good sense to recognize this phenomenon as a species of moral distress. It is distress that arises from a sense of being morally compromised, and it contributes to the sorts of practical problems traditionally associated with moral distress.

4.2   A New Definition of Moral Distress

We have argued that our understanding of moral distress should make space for the six types of cases we have discussed. But how should we revise our understanding of moral distress to encompass these cases? What definition of moral distress should we accept? There are countless possibilities, but we offer the following as a promising candidate:

> *Moral distress* $=_{df}$ one or more negative self-directed emotions or attitudes that arise in response to one's perceived involvement in a situation that one perceives to be morally undesirable.

This definition of moral distress has some elements that require clarification. First, it implies that moral distress is a matter of having negative emotions or attitudes that are *self-directed*. These might include self-criticism, guilt, shame, embarrassment, lowered self-esteem, or anger toward oneself or about one's behavior.[7] The restriction to self-directed attitudes is meant to rule out cases in which a person has only other-regarding negative emotions in response to being involved in a morally undesirable situation. Suppose, for instance, that a nurse feels resentment and anger toward a doctor for involving him in the morally questionable treatment of someone who is not his patient. He might feel angry about being involved with this case without feeling that *he* is morally compromised by the involvement. If moral compromise is at the heart of moral distress, it seems essential that there are self-directed negative emotions.

Our definition concerns one's perceived *involvement*. This is intentionally vague, allowing for a wide range of ways in which individuals might be related to a morally

---

[7] These attitudes should spring from a certain appreciation of moral values and not a purely instrumental concern. An egoist or a psychopath might strive to avoid acting immorally solely because it can bring about unwelcome legal and social consequences. If he slips up and does something wrong, he might chastise himself for his stupidity and carelessness. This would not be moral distress.

problematic situation. The involvement might be a matter of having acted or failed to act in certain ways, or having felt or failed to feel certain things. It might be that one has oversight over, and responsibility for, a situation even if one is in no position to intervene. Or it might be that one is simply connected, professionally or personally, to others who are more centrally involved in a morally undesirable situation. Since moral distress is grounded in individuals' perceptions of their involvement, and since individuals will vary quite a bit in the levels and types of involvement that lead them to feel morally compromised, it is ideal for a definition of moral distress to leave space for this variation.

The proposed definition of moral distress refers to situations perceived to be *morally undesirable*. It is notoriously difficult to define "moral." For our purposes, we understand morality to be concerned with the concern and respect that is owed to others. What types of being count as morally relevant "others" (or, we might say, beings with moral status) is a matter of dispute. Variation in people's views about the moral status of a given type of subject—for example, fetuses, animals, brain-dead patients—can help to explain variation in their experience of moral distress. Situations are morally desirable to the extent that due concern and respect to others are shown, and morally undesirable to the extent that they are not. The notion of a morally undesirable situation is meant to be somewhat open-ended. It might include situations that are morally optimal but still morally bad (e.g., where one chooses the lesser of two evils), as well as situations that are morally good but morally nonoptimal (e.g., where one chooses the lesser of two moral goods).

There are some notable contrasts between our proposed definition and the traditional understanding of moral distress. On the traditional view, moral distress is restricted to situations in which, due to constraints, one fails to do what one takes to be the morally right thing. However, cases of moral uncertainty reveal that the restriction to knowledge is too strong. Cases of bad moral luck suggest that moral distress can result from doing what was, in light of the information available at the time, the morally right thing. Cases of moral dilemma show that there need not be a "morally right thing to do." Cases of distress by association show that an individual's own action or omission need not be the source of distress—and, in turn, that the presence of internal or external constraints on action is not essential to moral distress. On our broader definition, moral distress can arise in situations where a person perceives herself to be involved in a morally undesirable situation. This allows for the possibility that an individual does not know what the morally right thing to do is (moral uncertainty), that the individual did the morally best thing though things turned out badly (bad moral luck), that there may not be a morally right thing to do (moral dilemma), or that one's own action is not the issue (distress by association).

Our definition does not place limitations on whether distress occurs at the very moment of one's involvement in a morally undesirable situation, afterward, or both. Unlike the traditional understanding of moral distress, which sees both initial distress and reactive distress as essential elements, our definition allows for the possibility that one does not have one of these. It therefore allows for cases of delayed distress, as well as cases in which one does not experience reactive distress or moral residue (which might happen in cases of mild distress). Interestingly, our definition even allows for the possibility of *anticipatory distress*. If a health care worker believes that, in the future, he or she *will* be involved in a morally undesirable situation, this can lead to distress in the present. This phenomenon may not be all that uncommon. Health care workers who have routinely found themselves entangled in morally undesirable situations can reasonably assume that they will find themselves in such situations in the future.[8] This can be a source of distress in their lives.

Lastly, the traditional understanding of moral distress implies that moral distress only arises from serious violations of one's values and therefore does not acknowledge instances of mild moral distress that, even if not terribly important on their own, can have a significant cumulative impact. In contrast, on our definition even mild forms of negative emotions and attitudes could constitute moral distress.

4.3   Is It Too Broad?

It might be thought that seeking a broader characterization of moral distress is a misguided goal. As we ourselves acknowledged in the first section, moral distress has gained such attention over the past several years primarily because it is a serious problem in actual health care practice. Bioethicists and health care practitioners want to understand what moral distress is in order to identify and remedy it in real-life contexts. However, our broader definition might seem to thwart that goal. Just think of the wide range of negative self-directed emotions individuals might feel, or the innumerable ways in which an individual might perceive herself to be "involved" in a situation, or the countless ways in which a situation might be judged to be morally undesirable. As Joan McCarthy and Rick Deady once observed, we do not want a definition of moral distress to be "so broad … as to be diagnostically and analytically meaningless" [10, p. 259].

This is a reasonable worry, but we think it admits of a satisfying response. Although it might prove difficult to conduct research on moral distress in general when it is so broadly conceived, it seems perfectly possible for researchers to overcome this problem by specifying a particular type of moral distress and making that their object of study. Could such a strategy prove fruitful? Thankfully, we need not resort to mere speculation here, for there is a prominent concrete case that sheds light on this question. The extant literature on moral distress is itself an in-depth investigation of one narrow (and important) type of moral distress—namely, moral distress that (1) results from the perception that one failed, due to internal and external constraints, to behave in the morally right way, (2) in a way that represents a compromising of one's moral integrity or core values, and that (3) involves both initial distress and reactive distress. Feature (1) represents one way in which an individual can perceive herself to be involved in a morally undesirable situation. Feature (2) will tend to involve or be correlated with very intense negative self-directed emotions and attitudes. Feature (3) concerns the time at which the distress

---

[8] It is conceivable that anticipatory distress will play some role in the best explanation of the so-called "crescendo effect." See Epstein and Hamric [11].

is experienced. If the existing research on moral distress represents a worthwhile endeavor (as we believe it has), then it is clear that investigating a particular type of moral distress can be worthwhile. Working with a broader understanding of moral distress is no impediment to focusing our research and interventions on narrower, context-pertinent forms of moral distress. In fact, our proposed definition of moral distress can support the investigation of particular forms of moral distress insofar as it lends itself to developing a taxonomy of moral distress, which serves to illuminate the full range of varieties of moral distress.

4.4   Toward a Taxonomy of Moral Distress

Given our broad definition, a taxonomy may be organized around three components of moral distress: the negative attitudes that one experiences, one's perceived involvement in the situation, and the perceived moral undesirability of the situation. While it is beyond the scope of this essay to work out this taxonomy in detail, the following is a rough sketch of the general form it might take and the practical, conceptual, and empirical implications:

The Negative Attitudes
  • The type of negative attitudes.
  • The appropriateness or fittingness of the attitudes.
  • The intensity of the attitudes.
  • The time at which the attitudes occur.
  • The positive and/or negative consequences of the attitudes (e.g., on one's job satisfaction, job performance, personal life).
The Perceived Involvement
  • The type of involvement.
  • The degree of involvement.
  • The accuracy of the perception of involvement.
The Perceived Moral Undesirability
  • The source of moral undesirability (i.e., what makes the situation morally undesirable).
  • The degree of moral undesirability.
  • The accuracy of the perception of moral undesirability.

In a fully developed taxonomy, each of the subcategories will be attached to a list of options. For example, the time at which the attitudes occur would include the following: before the time of one's perceived involvement, during the time of one's perceived involvement, after the time of one's perceived involvement, or some combination of these. For any given instance of moral distress, it can be asked how it should be classified within each subcategory.

Developing a taxonomy of moral distress can be beneficial. From a practical and conceptual perspective, it can open our eyes to the many varieties of moral distress, preventing us from becoming narrowly focused on a particular type that does not have moral priority over various other types. Empirically, it can also stimulate new lines of thinking about how to deal with moral distress by examining, for instance, the relationship between clinicians' degree of involvement in morally distressing situations, the source and degree of the moral undesirability of these situations, possible mitigation strategies, and clinician- and patient-related outcomes. To illustrate, consider this question: Should a hospital seek to prevent the occurrence of situations that will be perceived as morally undesirable by its staff, should it seek to ameliorate the moral distress experienced by the staff after such situations occur, or should it attempt to do both? Reflecting on the outlined taxonomy, it is evident that the choice between different interventions will depend crucially on the type of moral distress in play. If the kind of situation that gives rise to moral distress involves a violation of patients' rights or serious harm to them, clearly there should be efforts to prevent such morally undesirable situations from occurring. However, suppose instead that we are focusing on moral distress that arises from patients or their families making cool-headed, informed, and legally protected medical decisions that the medical staff considers to be foolish or immoral. With moral distress of this kind, it seems far more plausible that the hospital should focus its effort on mitigating moral distress in its staff without trying to prevent the occurrence of the situations that give rise to that distress. Thus, different forms of moral distress will call for different types of responses and interventions.

Importantly, we are not claiming that all forms of moral distress require or merit intervention. The envisioned taxonomy of moral distress will reveal instances of moral distress that are plausibly best left to individuals to address on their own. These might include cases where the moral distress is mild, the distress springs from obviously misguided moral views or unreasonable beliefs about one's involvement, the intensity of moral distress is much greater than the situation warrants, or individuals are having normal and appropriate self-critical responses to their own moral failings. Working out a taxonomy of moral distress can help us systematically explore whether and how we should intervene to address moral distress.

4.5   Conclusion

In this essay, we have sought to motivate the need for a broader definition of moral distress, propose a broader definition, and gesture toward a taxonomy that might be developed from this definition. While the appeal of our proposed definition partly depends upon the success of our case for favoring a broader understanding of moral distress (presented in the first section), the success of our case for a broader definition does not depend on the appeal of our definition. Thus, some readers might be convinced by the considerations in the first section and yet find some reason to reject the definition presented in the second section. We welcome this. If our proposed definition of moral distress proves to be unacceptable for reasons we have not foreseen, we hope that our attempt will inspire others to discover a better one.

Disclaimer The views expressed in this essay are the authors' own. They do not represent the positions or policies of the National Institutes of Health, U.S. Public Health Service, or Department of Health and Human Services.

The article was originally published in the American Journal of Bioethics, 16(12): 2–9, 2016.

Acknowledgments For their helpful feedback on an earlier version, we thank Ryan Antiel, Vanessa Carbonell, Nick Evans, Moti Gorin, Chris Feudtner, Steven Joffe, Sven Nyholm, David Wasserman, and Mindy Zeitzer. In addition, we thank an anonymous reviewer for drawing our attention to Carina Fourie's recent essay "Moral Distress and Moral Conflict in Clinical Ethics" (2015) [25], which also seeks to motivate a broader (though different) definition of moral distress and makes similar points regarding moral uncertainty and moral dilemmas. We recommend her essay to those interested in this topic.

Suggested Readings

American Association of Critical-Care Nurses. Moral distress position statement. Aliso Viejo: AACN; 2008.

Austin W, Kelecevic J, Goble E, Mekechuk J. An overview of moral distress and the paediatric intensive care team. Nurs Ethics. 2009;16(1):57–68.

Canadian Nurses Association. Code of ethics for registered nurses. Ottawa: CNA; 2008.

Chen PW. When nurses and doctors can't do the right thing. New York Times. 2009, February 5. Available at: www.nytimes.com/2009/02/06/health/05chen.html.

Corley MC. Nurse moral distress: A proposed theory and research agenda. Nurs Ethics. 2002;9:636–50.

Corley MC, Elswick RK, Gorman M, Clor T. Development and evaluation of a moral distress scale. J Adv Nurs. 2001;33(2):250–6.

Epstein EG, Delgado S. Understanding and addressing moral distress. Online J Issues Nurs. 2010;15(3):1–12.

Epstein EG, Hamric AB. Moral distress, moral residue, and the crescendo effect. J Clin Ethics. 2009;20(4):330–42.

Fourie C. Moral distress and moral conflict in clinical ethics. Bioethics. 2015;29:91–7.

Hamric AB. A case study of moral distress. J Hosp Palliat Care Nurs. 2014;16(8):457–63.

Hamric AB, Blackhall LJ. Nurse-physician perspectives on the care of dying patients in intensive care units: collaboration, moral distress, and ethical climate. Crit Care Med. 2007;35(2):422–9.

Hamric AB, Borchers CT, Epstein EG. Development and testing of an instrument to measure moral distress in healthcare professionals. AJOB Prim Res. 2012;3(2):1–9.

Hanna DR. Moral distress: the state of the science. Res Theory Nurs Pract. 2004;18:73–93.

Jameton A. Nursing practice: the ethical issues. Englewood Cliffs: Prentice Hall; 1984.

Jameton A. Dilemmas of moral distress: moral responsibility and nursing practice. AWHONNS Clin Issues Perinat Womens Health Nurs. 1993;4:542–51.

Kelly B. Preserving moral integrity: a follow-up study with new graduate nurses. J Adv Nurs. 1998;28:1134–45.

McCarthy J, Deady R. Moral distress reconsidered. Nurs Ethics. 2008;15(2):254–62.

Oshana M. Moral taint. Metaphilosophy. 2006;37(3–4):353–75.

Rushton CH. Defining and addressing moral distress. ACCN Adv Crit Care. 2006;17(2):161–8.

Ulrich CM, Hamric A, Grady C. Moral distress: a growing problem in the health professions? Hastings Cent Rep. 2010;40(1):20–2.

Webster GC, Baylis F. Moral residue. In: Rubin SB, Zoloth L, editors. Margin of error: the ethics of mistakes in the practice of medicine. Hagerstown: University Publishing; 2000. p. 217–30.

Whitehead PB, Herbertson RK, Hamric AB, Epstein EG, Fisher JM. Moral distress among health-care professionals: report of an institution-wide survey. J Nurs Scholarsh. 2015;47(2):117–25.

Wilkinson JM. Moral distress in nursing practice: experience and effect. Nurs Forum. 1987–1988;23(1):16–29.

Williams B. Ethical consistency. Proc Aristot Soc Suppl. 1965;39:103–24.

Williams B, Nagel T. Moral luck. Proc Aristot Soc. 1976;50:115–51.

4.6   A Broader Understanding of Moral Distress Revisited

Stephen M. Campbell, Connie M. Ulrich, and Christine Grady

Our essay "A Broader Understanding of Moral Distress" was published in the December 2016 issue of *American Journal of Bioethics* alongside a guest editorial and twelve commentaries from colleagues working in bioethics, medicine, nursing, and philosophy. The responses were largely favorable. The guest editorial—written by two physician-ethicists from Stanford—welcomed the expansion of the moral distress concept to make space for "the lesser known, more nuanced relatives" of traditional moral distress that are "not quite destructive to moral integrity and not intractable in the situation, but unsettling enough that they deserve thoughtful attention, exploration and, when possible, mediation and resolution" [26, p. 2].

A majority of the commentators were also sympathetic to, if not persuaded by, our case for broadening the definition of moral distress in order to make it more inclusive. Many of these same authors took the opportunity to explore interesting dimensions of this topic. Stephen Latham [27] highlighted parallels between our definition and the Catholic doctrine of complicity with wrongdoing. Andrew McAninch [28] drew connections between the concepts of moral distress and moral injury and explored the implications of recognizing distress in cases of luck. Markus Christen and Johannes Katsarov [29] thoughtfully examined the relationship between moral distress and moral sensitivity. David Resnik [30] offered some pioneering reflections on the presence of moral distress in the context of scientific research. Sven Nyholm [31] convincingly argued that there is much good in a person's propensity to experience appropriate moral distress and that we should not lose sight of this fact. Carolyn W. April and Michael D. April [32] brought Rawlsian considerations to bear on our discussion and highlighted practical advantages of our approach. These contributions have enriched our own understanding of various facets of moral distress.

However, some respondents were critical of our proposal, and we would like to briefly address what we regard as the two most important objections to our proposal. The first comes from Epstein et al. [33]. These authors raised worries about the practical effects of abandoning the traditional understanding of moral distress in favor of a broader definition. In particular, their concern was that a broader definition "dilutes the concept to such a degree as to render it impractical—too nebulous to be effectively taught, studied, used in practice, or, frankly, respected any longer as a powerful phenomenon in bioethics." This is an understandable concern, but to repeat a point that was explicitly presented in our essay and merits repeating here, our broader definition of moral distress is in no way incompatible with recognizing and researching specific context-pertinent forms of moral distress. Indeed, our definition with its accompanying proposed taxonomy facilitates the identification of specific forms of moral distress and positions us to better understand and appreciate their various dimensions and to assess their ethical significance. Hence, the authors' emphasis on the importance of "naming the moral distress experienced by staff" seems misplaced. Our definition is no obstacle to telling people, "You're experiencing a kind of moral distress." (In fact, this is precisely what should be said even on the traditional narrow definition since that definition also admits of different subspecies of moral distress.) Furthermore, as Carina Fourie [34] has helpfully suggested, it is possible that there are ways to justify the special importance of the traditional definition's "constraint-distress" in nursing contexts. If its special importance can be justified, our broader definition should not be so worrisome. Finally, we feel that this line of criticism misses the immense practical benefit and importance of recognizing a wider array of relevantly similar phenomena.

The second important line of objection was presented in the commentary from Moti Gorin [35]. Gorin introduced the following hypothetical case:

Sexual Harassment: A female nurse enters the break room for coffee. Two of her male colleagues are sitting at a table and eating. As the woman is leaving with her coffee, one of the men makes a lewd comment to his lunch partner about the appearance of their female coworker. He makes the comment openly, clearly with the intention that she will hear it. The other man responds with laughter. As she exits the room the woman is overcome with feelings of annoyance, anger, and fear, which are directed at her colleagues. She also feels acute shame and embarrassment as a result of being crassly objectified. Even when the passage of time has reduced the intensity of these emotions, she can't help feeling less confident whenever she's at the hospital.

Gorin's case reveals a flaw in our proposed definition. On the one hand, it does not seem intuitively correct to say that the woman in this case experiences *moral* distress or that she herself feels *morally* compromised or tainted by this interaction. She can walk away from the exchange with a clear conscience, despite whatever other feelings she may have. On the other hand, her emotional response does qualify as moral distress on our definition, which identifies moral distress with "one or more negative self-directed emotions or attitudes that arise in response to one's perceived involvement in a situation that one perceives to be morally undesirable" Campbell et al. [37, p. 6]. In Gorin's imagined case, the woman does indeed have negative self-directed emotions (shame, embarrassment) in response to her involvement (victim) in a situation that she perceives to be morally undesirable (an instance of sexual harassment). We concede that this is a case where our proposed definition overgenerates or, we might say, delivers a false positive.

The question is: how do we adjust the definition to handle this problem? Gorin rightly notes that the problem has to do with our underdeveloped notion of "involvement." He suggests that we modify our definition by replacing "perceived involvement in" a morally undesirable situation with "perceived [moral] responsibility for" a situation thought to be morally undesirable. This would avoid the implication that the woman in his case has moral distress provided that she does not perceive herself as being morally responsible for the harassment she suffers. We agree with Gorin that one's involvement needs to have a moral dimension, but we also think that a modification phrased in terms of responsibility is too strong. There are cases of bad moral luck and distress by association where a person might be fully convinced that she is *not* morally responsible for the morally undesirable situation with which she is involved and yet still feels morally compromised by her association with the causal consequences of her actions or with other parties. Our preferred solution is to modify our definition so that one who has moral distress must perceive her involvement in a morally undesirable situation to be itself morally undesirable:

> Moral distress = one or more negative self-directed emotions or attitudes that arise in response to one's perceived morally undesirable involvement in a situation that one perceives to be morally undesirable.

This modification retains the spirit of our original proposal and addresses Gorin's objection by screening out cases where one has a morally unproblematic involvement in a morally problematic situation. We are optimistic that our definition can be successfully tweaked to handle new counterexamples that emerge.

Moral distress is a global phenomenon that is widely experienced by those working in health care. Our essay outlining six example cases was meant to broaden the dialogue on this valuable and pervasive problem. We are pleased that our proposed definition of moral distress welcomed such an engaging dialogue on what it is and what it is not. Future work is now needed to develop a taxonomy around the three critical components of our definition: the negative attitudes that one experiences, one's perceived involvement, and the perceived moral undesirability of the situation. This type of work is bound to open up new areas of normative and empirical bioethics research, further refining the depth, dimensions, and significance of this important topic for all those who face the everyday ethical challenges of caring for patients and families.[9]

---

[9] This essay is adapted from Campbell et al. [36].

Suggested Readings

April CW, April MD. Understanding moral distress through the lens of social reflective equilibrium. Am J Bioeth. 2016;16(12):25–7.

Burgart AM, Kruse KE. Moral distress in clinical ethics: expanding the concept. Am J Bioeth. 2016;16(12):1.

Campbell SM, Ulrich CM, Grady C. A broader understanding of moral distress. Am J Bioeth. 2016;16(12):2–9.

Campbell SM, Ulrich CM, Grady C. Response to open peer commentaries on 'A broader understanding of moral distress'. Am J Bioeth. 2016;16(12):W1–3.

Christen M, Katsarov J. Moral sensitivity as a precondition of moral distress. Am J Bioeth. 2016;16(12):19–21.

Epstein E, Hurst A, Mahanes S, Marshall MF, Hamric A. Is broader better? Am J Bioeth. 2016;16(12):15–7.

Fourie C. The ethical significance of moral distress: inequality and nurses' constraint-distress. Am J Bioeth. 2016;16(12):23–5.

Gorin M. The role of responsibility in moral distress. Am J Bioeth. 2016;16(12):10–1.

Latham B. Moral distress and cooperation with wrongdoing. Am J Bioeth. 2016;16(12):31–2.

McAninch A. Moral distress, moral injury, and moral luck. Am J Bioeth. 2016;16(12):29–31.

Nyholm S. The normative and evaluative status of moral distress in health care contexts. Am J Bioeth. 2016;16(12):17–9.

Resnik D. Moral distress in scientific research. Am J Bioeth. 2016;16(12):13–5.

References

A Broader Understanding of Moral Distress

1. Jameton A. Nursing practice: the ethical issues. Englewood Cliffs: Prentice Hall; 1984.
2. Wilkinson JM. Moral distress in nursing practice: experience and effect. Nurs Forum. 1987–1988;23(1):16–29.
3. Jameton A. Dilemmas of moral distress: moral responsibility and nursing practice. AWHONNS Clin Issues Perinat Womens Health Nurs. 1993;4:542–51.
4. Webster GC, Baylis F. Moral residue. In: Rubin SB, Zoloth L, editors. Margin of error: the ethics of mistakes in the practice of medicine. Hagerstown: University Publishing; 2000. p. 217–30.
5. Corley MC. Nurse moral distress: A proposed theory and research agenda. Nurs Ethics. 2002;9:636–50.
6. Hanna DR. Moral distress: the state of the science. Res Theory Nurs Pract. 2004;18:73–93.
7. Rushton CH. Defining and addressing moral distress. ACCN Adv Crit Care. 2006;17(2):161–8.
8. American Association of Critical-Care Nurses. Moral distress position statement. Aliso Viejo: AACN; 2008.
9. Canadian Nurses Association. Code of ethics for registered nurses. Ottawa: CNA; 2008.
10. McCarthy J, Deady R. Moral distress reconsidered. Nurs Ethics. 2008;15(2):254–62.
11. Epstein EG, Hamric AB. Moral distress, moral residue, and the crescendo effect. J Clin Ethics. 2009;20(4):330–42.
12. Austin W, Kelecevic J, Goble E, Mekechuk J. An overview of moral distress and the paediatric intensive care team. Nurs Ethics. 2009;16(1):57–68.
13. Chen PW. When nurses and doctors can't do the right thing. New York Times. 2009, February 5. Available at: www.nytimes.com/2009/02/06/health/05chen.html.
14. Ulrich CM, Hamric A, Grady C. Moral distress: a growing problem in the health professions? Hastings Cent Rep. 2010;40(1):20–2.

15. Epstein EG, Delgado S. Understanding and addressing moral distress. Online J Issues Nurs. 2010;15(3):1–12.
16. Hamric AB. A case study of moral distress. J Hosp Palliat Care Nurs. 2014;16(8):457–63.
17. Corley MC, Elswick RK, Gorman M, Clor T. Development and evaluation of a moral distress scale. J Adv Nurs. 2001;33(2):250–6.
18. Hamric AB, Blackhall LJ. Nurse-physician perspectives on the care of dying patients in intensive care units: collaboration, moral distress, and ethical climate. Crit Care Med. 2007;35(2):422–9.
19. Hamric AB, Borchers CT, Epstein EG. Development and testing of an instrument to measure moral distress in healthcare professionals. AJOB Prim Res. 2012;3(2):1–9.
20. Kelly B. Preserving moral integrity: a follow-up study with new graduate nurses. J Adv Nurs. 1998;28:1134–45.
21. Whitehead PB, Herbertson RK, Hamric AB, Epstein EG, Fisher JM. Moral distress among healthcare professionals: report of an institution-wide survey. J Nurs Scholarsh. 2015;47(2):117–25.
22. Williams B. Ethical consistency. Proc Aristot Soc Suppl. 1965;39:103–24.
23. Williams B, Nagel T. Moral luck. Proc Aristot Soc. 1976;50:115–51.
24. Oshana M. Moral taint. Metaphilosophy. 2006;37(3–4):353–75.
25. Fourie C. Moral distress and moral conflict in clinical ethics. Bioethics. 2015;29:91–7.

A Broader Understanding of Moral Distress Revisited

26. Burgart AM, Kruse KE. Moral distress in clinical ethics: expanding the concept. Am J Bioeth. 2016;16(12):1.
27. Latham B. Moral distress and cooperation with wrongdoing. Am J Bioeth. 2016;16(12):31–2.
28. McAninch A. Moral distress, moral injury, and moral luck. Am J Bioeth. 2016;16(12):29–31.
29. Christen M, Katsarov J. Moral sensitivity as a precondition of moral distress. Am J Bioeth. 2016;16(12):19–21.
30. Resnik D. Moral distress in scientific research. Am J Bioeth. 2016;16(12):13–5.
31. Nyholm S. The normative and evaluative status of moral distress in health care contexts. Am J Bioeth. 2016;16(12):17–9.
32. April CW, April MD. Understanding moral distress through the lens of social reflective equilibrium. Am J Bioeth. 2016;16(12):25–7.
33. Epstein E, Hurst A, Mahanes S, Marshall MF, Hamric A. Is broader better? Am J Bioeth. 2016;16(12):15–7.
34. Fourie C. The ethical significance of moral distress: inequality and nurses' constraint-distress. Am J Bioeth. 2016;16(12):23–5.
35. Gorin M. The role of responsibility in moral distress. Am J Bioeth. 2016;16(12):10–1.
36. Campbell SM, Ulrich CM, Grady C. Response to open peer commentaries on 'A broader understanding of moral distress'. Am J Bioeth. 2016;16(12):W1–3.
37. Campbell SM, Ulrich CM, Grady C. A broader understanding of moral distress. Am J Bioeth. 2016;16(12):2–9.

Sources of Moral Distress

Mary K. Walton

5.1 Introduction

Since Andrew Jameton [1] defined the concept of moral distress over three decades ago, clinicians and ethicists have sought to define the phenomenon and understand its impact on professionals, patients, and organizations. The experience of moral distress is associated with negative consequences for both people and health-care systems, including clinician burnout and poor patient outcomes [2–4]. Identifying the root causes of moral distress, distinguished from other emotional stressors inherent in health care [5], is necessary to develop and study strategies to mitigate or prevent its negative consequences. First identified as a phenomenon nurses experienced in acute care practice, moral distress is now known to be experienced in a wide variety of clinical settings and by all professional groups [6–9] and is recognized as a significant source of moral suffering among nurses and other clinicians. Descriptive studies have revealed root causes of moral distress that extend beyond Jameton's [1] focus on institutional constraints and power hierarchies [10].

As researchers have studied the phenomenon in other disciplines and practice norms have evolved, additional root causes have been identified through qualitative, quantitative, and mixed methods studies. Clinicians and researchers continue to explore the phenomenon [11], proposing expanded definitions [12] and revealing internal factors, clinical practice patterns, and cultural norms and changing professional roles that contribute to our evolving understanding of this significant phenomenon. The relational aspect of moral distress is essential to recognize; although experienced by individuals and shaped by their moral characteristics, it is also shaped by "the multiple contexts within which the individual is operating, including

M.K. Walton
Department of Nursing, Hospital of the University of Pennsylvania,
Philadelphia, PA, USA
e-mail: Mary.Walton@uphs.upenn.edu

© Springer International Publishing AG 2018
C.M. Ulrich, C. Grady (eds.), *Moral Distress in the Health Professions*,
https://doi.org/10.1007/978-3-319-64626-8_5

the immediate interpersonal context, the health care environment and the wider socio-political and cultural context" [13]. As this interrelationship extends beyond the health-care practice setting, it is imperative to understand the sources of moral distress and address the negative consequences for professionals and organizations toward the goal of achieving the best possible health outcomes for patients and communities.

5.2   Major Root Causes of Moral Distress

Researchers organize the root causes for moral distress into three broad categories: clinical situations, internal constraints, and external constraints [10] (refer to Table 5.1). Although this broad categorization is useful, given the complex nature of health care, these categories may overlap, are interrelated, and may not be comprehensive. Further, other causes will likely be identified in the future. Scientific advances will introduce new treatment options, and care will be delivered by inter-professional teams to patients who will be expected to be increasingly engaged in

Table 5.1 Major root causes of moral distress

Clinical situations	
• Providing unnecessary/futile treatment	• Using resources inappropriately
• Prolonging the dying process through aggressive treatment	• Providing care that is not in the best interest of the patient
• Inadequate informed consent	• Providing inadequate pain relief
• Working with caregivers who are not as competent as care requires	• Providing false hope to patients and families
• Lack of consensus re-treatment plan	• Hastening the dying process
• Lack of continuity of care	• Lack of truth-telling
• Conflicting duties	• Disregard for patient wishes
Internal constraints	
• Perceived powerlessness	• Lack of knowledge of alternative treatment plans
• Inability to identify the ethical issues	• Increased moral sensitivity
• Lack of understanding the full situation	• Lack of assertiveness
• Self-doubt	• Socialization to follow others
External constraints	
• Inadequate communication among team members	• Tolerance of disruptive and abusive behavior
• Differing inter- (e.g., RN to MD) or intra-professional (e.g., RN to RN) perspectives	• Compromising care due to pressure to reduce costs
• Inadequate staffing and increased turnover	• Hierarchies within health-care system
• Lack of administrative support	• Lack of collegial relationships
• Policies and priorities that conflict with care needs	• Nurses not involved in decision-making
• Following family wishes of patient care for fear of litigation	• Compromised care due to insurance pressure or fear of litigation

Hamric, Borchers, Epstein. AJOB Primary Research, 2012: 3(2) p. 2

their own health care. All this will unfold within our complex and pluralistic society. Identifying and understanding the root causes of moral distress emerging from the ever-evolving health-care environment will continue to be a challenge for clinicians, administrators, and researchers alike.

Not all distress experienced by clinicians is moral distress. Cribb [14] characterizes the routine and pervasive nature of the burden inherent in health-care professional's work as moral stress and posits an ethical duty to accommodate some level of this stress. Berlinger's in-depth examination of the moral problems within health-care systems explores the routine tensions and dilemmas arising from clinicians' intimate contact with the problems of humanity and exposure to the realities of mortality [15]. This routine exposure for individuals and teams elicits emotions that should place moral obligations on the health-care organization or system to prevent problems for workers and teams. Epstein and Hurst [16] claim that moral distress is fundamentally a grave organizational problem albeit experienced on a personal level. Sources of moral distress reflect "the real and inescapable moral complexity of many clinical situations, including the fact that conscientious and thoughtful clinicians, patients and families can struggle with uncertainty, feel constrained by the pressures and limitations of time and resources, and disagree about ethically appropriate interventions and optimal outcomes" [17].

5.2.1 Clinical Situations

New options for care available in intensive care settings as well as the values and patterns of consistently providing aggressive care to critically ill individuals resulted in nurses questioning their practice and at times experiencing moral distress [18–21]. Nursing practice in intensive care units was foundational to the identification and early study of the phenomenon of moral distress. Now novel and invasive therapies are offered along the continuum of care whether in the intensive care unit or in the home; and interprofessional expertise and the support of family caregivers are essential to its provision. Over the past two decades researchers have documented moral distress in a multitude of practice settings [22]. Studies focusing on nurses describe moral distress in critical care [7, 23], medical-surgical units [24], schools [25], emergency departments [26, 27], and trauma care [28]. This author's practice includes working with home care nurses experiencing moral distress when they question the impact of aggressive care for the patient or are unable to provide for patient needs due to the sociopolitical context (i.e., access to preventative care, mental health services, adequate housing, and chronic home care needs). Moral distress may also emerge in undergraduate nursing educational experiences [29]. With the increasing recognition of the importance of interprofessional practice, other studies document moral distress in medical students [30], internal medicine trainees [31, 32], as well as multiple professional health-care disciplines [7, 8]. Moral distress is linked to the presence of some kind of constraint on moral agency [33]. Moral agency, the capacity to habitually act in an ethical manner, is relevant in all aspects of practice as ethical considerations are embedded in the countless everyday interactions in health-care settings [34–38]. Moral agency is not "merely the possession and adequacy of the agent's moral intentions or character. Moral agency is also based on experienced-based moral perception in practical situations and the nurse's capacity to respond quickly and effectively" [39]. Therefore, there is the potential for moral distress to be experienced wherever and whenever professionals recognize their obligation to assess, plan, predict, and control their provision of care.

5.2.1.1 Technology

Modern bioethics was in part a response to ethical questions arising from the growth in scientific knowledge and the accompanying technological advances that generated myriad treatment options. These advances have been embraced and integrated into practice with enthusiasm as professionals witnessed dramatic improvements in the care these technologic innovations made possible. Frontline nurses and physicians support both the critically and chronically ill and recognize that patients live longer and often better despite dependency on treatment interventions that technology makes possible. However, many of the determinants of moral distress arise through the use of technology in ways that professionals believe does not benefit the patient or provide value to organizations or society.

The ethical as well as legal value placed on the informed consent process focuses clinicians' attention on fulfilling their obligation to the process when involved with offering, initiating, providing, and discontinuing care deemed by them to be unnecessary, inappropriate, or futile. Lack of consensus regarding the treatment plan, care that only prolongs the dying process, care that only gives false hope, care believed to not be in the best interest of the patient, and using resources inappropriately have all been identified as root causes of moral distress embedded in clinical practice. Clinicians and researchers alike recognize the need to improve communication between health-care professionals and patients living with serious illness [40]. How should providers introduce therapies given the variation in health literary, the complexity of options, and the diversity of values and beliefs that inform these decisions on the part of all stakeholders? Technology introduces questions and options for all health-care workers along with patients and their families to consider. Ethical, medical, and legal considerations are embedded in considering the range of treatment options and determining realistic goals of care when there is serious illness. The adequacy of the informed consent process raises concerns given the impact of these technologies on quality of life considerations, considerations relevant both to the patient and their family caregivers. Although intractable disagreements in critical care settings may be the most visible, clinicians and patients alike grapple with myriad treatment options or alternately the lack of access to options in settings such as ambulatory clinics, rehabilitation centers, and school-based programs. Furthermore the impact of managing many of these interventions places a significant burden on family caregivers when they are continued in the home environment; there are societal consequences as caregivers balance other family responsibilities. Substantial evidence indicates that family caregivers of older adults have higher rates of depressive symptoms, anxiety, stress, and emotional difficulties than non-caregivers; research also shows caregivers are at risk for economic harm [41]. In addition to assisting with activities of daily living, nearly half of family caregivers report responsibility for complex medical tasks that are typically the province of a professional nurse or trained technician [42]. Technological advances fuel many situations that create moral distress for health professionals particularly when they must provide therapies they deem negatively impact their patient's quality of life rather than sustain or improve it. Recognition of the societal impact is not invisible to them as well.

5.2.1.2 Care Near the End of Life

Moral distress has been frequently described by health-care professionals caring for patients at the end of life when continuing aggressive care is the focus rather than providing for comfort and a peaceful death [43, 44]. In critical care settings, similar situations evoked moral distress in physicians and nurses, with the most distressing situations involved feeling pressured to continue aggressive treatment in situations where they did not believe it was warranted [45]. When death is approaching, those close to the individual will likely suffer; this proximity to suffering carries a moral burden for the professionals witnessing the suffering of patients, family, and friends. Whether an acute crisis or the end of a long journey with chronic health problems, the options for cure or extension of life will at some point be exhausted. The recent Institute of Medicine report, Dying in America [46], highlights the poor quality of communication between clinicians and patients with advanced illness, "particularly with respect to discussing prognosis, dealing with emotional and spiritual concerns, and finding the right balance between hoping for the best and preparing for the worst" [46] p.12. When patients or families request "do everything" to stave off death, clinicians face an ethical quandary of providing care that may serve only to extend the dying process and may inflict unnecessary pain and suffering for all involved. The person who needs critical care may be physiologically unstable and at risk or in danger of dying; intensive care and aggressive treatment is usually provided in the expectation of hope no matter how slim. In fact, "critical care by definition melds physiological instability with hope for survival" [47], p. 3. Critical care professionals often struggle with family requests for care seen as "futile" through their clinician lens; these requests may compel them to act against the best understanding of their professional obligations [48]. Recommendations for conflict resolution and preventing moral distress focus on reframing discussions of life-extending care to distinguish patients' goals and family preferences in contrast to those of critical care medicine. Initiating conversations about goals of care early in a chronic illness and when the situation is less tumultuous than in the intensive care setting can help to calm fears as illness progresses. Additionally, training for clinicians across disciplines in value clarification, creation of moral spaces, and communication as well as intensive symptom and pain management and concomitant training and research are needed to make better, proactive, and preventative symptom management possible. Clinicians with well-honed communication skills are able to explore and understand patient or family values and care preferences; therefore, they are able to frame and offer options that align with those values rather than clinical ones. Moral distress may be mitigated or even prevented through discussion and shared decision-making that honors both patient and clinician values and obligations.

5.2.1.3 Interprofessional Practice

Moral concerns are grounded in an individual's personal and professional values. Beliefs about how best to proceed in ethically complex treatment situations will in part be influenced by past experiences and professional training. How any one individual sees a situation and determines how best to enact their professional obligations will vary. Medicine's well-established hierarchy and requirements for years of clinical training brings professionals with varied knowledge and skill sets as well as diverse beliefs and values to the bedside. Findings in a study focused on ICU intrateam dynamics suggest that discordance within a team is a prominent source of moral distress across health professions [49]. In academic settings, highly experienced nurses may be dependent on medical trainees with less knowledge and experience to obtain "orders" for needed care. Care that is deemed "futile" by one clinician may be seen as restorative or curative by another. Other circumstances that influence intra-professional relationships but may be invisible in the daily intensive work of an acute care setting include each person's guiding philosophies or beliefs, spiritual or religious beliefs, and cultural norms. These personal characteristics influence patterns of decision-making and assessments of acceptable benefits and burdens for patient suffering. The concept of a moral community emerged in one study where critical care nurses with strong communication and conflict resolution skills saw themselves as essential to the decision-making process regarding the withdrawing of life-sustaining treatment and described practicing in "supporting" relationship with physician colleagues; moral distress was not described [50]. Redefining and exploring this concept of a moral community where the work of professionals is characterized not as teamwork but rather moral work where the shared goal is the well-being of patients may offer an antidote to moral distress. Sources of moral distress arising from conflicting professional duties, lack of continuity of care and truth-telling, and disregard for patient wishes may be less likely to develop.

5.2.1.4 Hastening the Dying Process

Societal norms about terminally ill individuals' right to control the timing of their death are changing; legislation in some states permits professionals to provide aid in the dying process. Professionals may experience moral distress when patients or even families request help to hasten the dying process toward the goal of ending suffering. Although professionals hold diverse views about the permissibility of this and codes of ethics vary as well, these requests undoubtedly place physicians, nurses, and pharmacists in situations that may challenge their values and beliefs. Whether the individual clinician views such requests as aid in the dying process or assisting in suicide, moral distress may be experienced [51].

5.2.2 Internal Constraints

5.2.2.1 Individual

The experience of moral distress is anchored in the individual's moral compass, one's sensitivity to and appreciation for their professional obligations, goals, and sense of self. Professionals exercise their moral agency when they make judgments based on their notions of right and wrong and accept accountability for their actions based on this judgment. Every individual holds core beliefs, values, and concepts that inform their sense of the good. Nurses, for example, have beliefs about the badness of pain and the definition of a good death. Perspectives based on one's gender, race, and life experiences influence assessments and actions. The risk of values imposition in clinical practice is recognized even for clinicians engaged in ethics consultations [52]. Sensitivity to this risk, the inability to identify an issue as an ethical one, and lack of knowledge about values and goals other than one's own may be the foundation for moral distress.

5.2.2.2 Professional Socialization
Since articulation of the original definition of moral distress, professional socialization has progressed from rigid patterns with stereotypical gender roles and inhibited dialogue characterized as "the doctor-nurse game" [53] to the establishment of educational competencies for interprofessional collaborative practice. With an ethics framework anchored in both professional and interprofessional identities, these competencies highlight the need for education in communication and teamwork [54]. Development of interprofessional educational curricula is underway albeit in the early stages. Medical hierarchy within health-care institutional settings is still a powerful force influencing the ability of medical or nursing trainees or licensed professionals to speak up, question "orders," or risk exposing one's lack of knowledge. Promoting an ethical climate where every individual feels empowered to raise a concern to be addressed by the team supports the beneficent and safe care of all patients.

5.2.2.3 Perceived Powerlessness
Perceived powerlessness is commonly cited in morally distressing situations and may be influenced and perhaps driven by the clinician's position in the health-care system and on the team. This author recalls medical students' descriptions of the variation in the changing and unspoken rules with each attending physician service rotation; this rotation pattern in the academic setting can be disruptive to the medical plan of care. Clinical nurses' positions have long been characterized as "in the middle" between physicians and patients where patients consent and physicians give "orders" as well as between patients and families or administrators and physicians among others [55–57]. There are many roles in the training of health-care professionals, student, intern, resident, fellow, and preceptor to name a few—descriptors and roles abound. The ability to direct and control one's own practice is influenced by one's position in both the formal and informal chain of authority. Physicians have power over, and accountability for, many aspects of patient care by virtue of their place in the medical hierarchy as well as legal and regulatory requirements for medical orders (i.e., medications, invasive interventions). Nurses, medical trainees, therapists, clinical dieticians, social workers, chaplains, and pharmacists may at times be indeed powerless in the immediate moment and unable to provide care needed and wanted by a patient. Moral distress by definition is one possible outcome of a professional's inability to provide appropriate care as a result of the established power structure.

5.2.2.4 Ethics Knowledge

Foundational education in ethics varies among health professions. Once in the practice setting, applying biomedical and philosophical principles to achieve ethical decision-making is challenging. Access to ethics resources and the quality of those resources in organizations is variable [58, 59]. There is some limited evidence that education and training in ethics has a significant influence on confidence, use of ethics resources, and moral action of social workers and nurses [60] and may decrease moral distress [61]. Continuing education programs in practice settings may be practical and valuable for clinicians and can focus specifically on real-life ethical concerns. Findings from a study focused on moral distress and psychological empowerment in critical care nurses support greater ethics education in nursing to support articulation of ethical principles and application in a multidisciplinary care context. [43]. Health care is increasingly planned, provided, and evaluated by an interprofessional team; ethical practice is enacted by professionals who express personal and professional values and hold each other accountable. Individual clinicians should identify what is morally relevant in a particular situation, appreciate their personal perspective, and seek to understand the multiple perspectives and positions of their colleagues. "If we are to be morally intelligible to one another, we must sustain or renew our understanding of moral terms-of what it means to speak of respect, client well-being, fidelity, or obligation" [62] (p. 37). Ethical competence includes "cultivating a rich moral vocabulary" [3] (p. 116). As a nurse ethicist, this author has found that coaching frontline nurses and physicians to raise, frame, and discuss concerns using the language of ethics helps them to engage their colleagues in a richer discussion focused on some of the well-recognized sources of moral distress, suffering without benefit or inadequate informed consent as representative examples. When clinicians express concerns about goals of care and suffering by reflecting a personal preference rather than a professional obligation, the discussion may not delve into the obligations, standards of care, or patient values, remaining only at the level of personal preferences. A professional who is able to express concerns within an ethics framework is more likely to promote discussion that may serve to prevent or mitigate sources of moral distress. This individual may have the opportunity to explore alternative treatment plans, gain a deeper understanding of the situation as seen by colleagues, and garner respect. The resulting care may be more person rather than provider-centric as well.

Another example of the power of an ethics vocabulary in the practice setting relates to the significant challenge of maintaining patient privacy and confidentiality. In this author's practice, nurses consistently reference legal and risk management constraints due to the HIPPA standards, implying they negatively impact patient care. However, when coached to consider the professional obligation to protect the patient's right to privacy and confidentiality as long established in the code of ethics [63] and the genesis for this obligation, the focus shifts. The rationale for this ethical obligation may be discussed with the consequences for patients (in contrast to those for organizations) and at times virtue ethics, i.e., character and trustworthiness. Thus, a much richer exploration may develop encompassing how to enact the professional's role as well and how to understand the patient's perspective, consider their needs, and also be reminded of the regard and esteem with which the public holds professional nurses; the important legal restrictions required by HIPPA legislation is secondary to ethical practice. Using an ethics framework may prevent legal and risk management considerations from overshadowing or even rendering invisible professional ethical obligations.

5.2.3 External Constraints

The sources of moral distress categorized by researchers as external constraints relate to the clinical situation and the knowledge, skills, and attitudes of the individual professional. However, this categorization of external constraints ensures that the organizational and sociopolitical variables impacting moral distress are given due consideration.

5.2.3.1 Person- and Family-Centered Care

When the IOM [64] identified patient-centered care as one of the six core imperatives to improve quality and safety, a cultural shift in health care gained momentum. The concept of patient centeredness is one of partnership where the expertise of the clinicians joins with the values, goals, and care preferences of the patient in the decision-making process; the patient is less a recipient of care and more "a source of control and full partner" [65] (p. 123). Clinicians bring specialized knowledge and skill to the process although perhaps variable skill and practice with shared decision-making and other communication skills enabling them to elicit patient values and care preferences. Clinicians trained in a culture that prioritizes science and clinical expertise needed to shift from the medical model with the centrality of the physician to one where the individual patient has a greater role in care decisions [66]. By definition, the partnership essential to person-centered care requires the clinician's expertise and clinical judgment to be part of the decision-making process along with patient values and care preferences. When in response to patient or family demands, physicians and nurses provide care deemed not to be the standard of care and not in the patient's best interest, moral distress will likely be experienced. Acquiescing to patient or family demands that compromise professional standards is an inaccurate understanding of person-centered care and a significant source of moral distress. Continuing overly aggressive care to a dying patient was conceptualized by a critical care nurse as a patient's body without a soul and feeling the care provided "was like ventilating a corpse" [19] (p. 232). Initiating resuscitation efforts for dying patients prompted a former critical care nurse to muse to this author that she often felt yellow crime scene tape should encircle the code cart as the teams' efforts felt criminal rather than caring.

© 2017 Anita F. McGinn-Natali

5.2.3.2 Following Family Wishes for Patient Care for Fear of Litigation

In addition to the mandate for yet inaccurate conceptualization of person-centered care, professionals fear litigation initiated by surviving family members when their demands for continued aggressive care are not honored. When patients are no longer able to direct their own care, family members and other loved ones often fill the role of surrogate decision-maker, whether appointed by the patient in a legal document or in accord with state legislation. When a loved one is near the end of life, shared decision-making is challenging [67], and families may reject recommendations to focus on comfort or withdraw life-sustaining therapy. Clinicians' ethical concerns can be overshadowed by legal and risk management considerations given the knowledge that the patient's death is inevitable, yet an angry family caught in an intractable dispute could initiate litigation or call attention to dissatisfaction with care through the media. Efforts continue to be directed toward conflict resolution strategies in these challenging situations.

5.2.3.3 Professionalism

Incivility, bullying, and disruptive behaviors that create hostile, unsafe work environments, perpetuate burnout and moral distress, and risk patient safety are pervasive and continue to receive attention [68–70]. Inadequate staffing patterns as a result of high levels of clinician turnover could be related to the tolerance of these behaviors, and research should address this concern. Professionals who are technically competent yet consistently demonstrate unprofessional behavior should not be considered competent professionals or "competent in a professionally comprehensive sense." Clinical and administrative leaders need to address this issue and establish programs for all stakeholders: perpetrators, victims, and bystanders. Training in communication including cognitive rehearsal for victims and bystanders and zero tolerance policies need to be enculturated into educational and practice settings alike.

5.2.3.4 Policies and Priorities

Internal organizational policies and priorities, as well as state legislation, can introduce barriers that block clinicians from providing care they believe they are obligated to provide. Regulatory standards and state legislation requires health-care organizations to have policies establishing a process for clinicians to opt out of providing care deemed to be in conflict with their moral, spiritual, or religious convictions, i.e., staff rights policies. Conscientious objection clauses provide professionals an opportunity to opt out of situations in which they find themselves morally compromised in some way; however, research on moral distress seems to indicate that perceived powerlessness is more common in practice than conscientious objection. Moral distress arises when clinicians believe they are morally and professionally obligated to provide care, and yet organizational policies and fiscal priorities block them from doing so, for example, reducing costs by reduction in services or personnel or refusing to individualize care to limit risk of liability. Physicians and nurses may choose to not participate in surgical procedures when a patient does not consent to blood transfusion or participate in the care of a woman undergoing an abortion by invoking conscientious objection; there is no comparable process to enable them to provide care when blocked by policies and priorities that conflict with care needs. Berlinger [71] calls on bioethicists to reclaim the idea of "conscience" and establish a framework for professionals to raise an objection when needed care is not provided. The narrow and negative notion of conscience defined only by refusal limits professionals' practice and creates the potential for moral distress. When professionals believe they are obligated to provide care yet are restricted from doing so, there may be no recourse to enable them to act based on their conscience.

5.2.3.5 Organizational Values and Ethical Climate

Organizational culture impacts the quality of life of those who practice; health system values, norms, and structures may support moral agency and integrity or promote moral distress and clinician burnout [72, 73]. Acknowledging and addressing the sources of stress and moral distress is increasingly recognized as an obligation of administrators, managers, and all those in leadership positions. This obligation is rooted in their leadership mandate to create a professional practice environment that creates and sustains a climate of respect in which safe and high quality care is provided [74–76]. Organizational norms established by administrators impact how the culture may perpetuate or mitigate moral stress in the practice environment. The pervasiveness of behaviors from incivility to lateral violence is well recognized. Resources as well as role modeling behaviors for a healthy work environment are essential; attention is now shifting to identify evidence-based interventions to mitigate moral distress and promote professional well-being and moral resilience [73, 76].

Administrators and clinical leaders must support institutional ethics resources beyond the norm of a single ethics committee that serves only to meet regulatory standards. Clinicians who bear the moral burden of proximity to human suffering while navigating the endless, everyday ethical concerns arising in modern health care need safe moral spaces to promote for self-care as well as for the provision high quality patient care [59].

5.2.3.6 Societal Factors

Inequalities in access to health care and health insurance place a significant burden on clinicians seeking to honor their personal values and professional obligations in the care of vulnerable individuals [77]. Today, about 11% of Americans still lack health insurance (https://www.cdc.gov/nchs/fastats/health-insurance.htm), and professionals throughout health-care systems globally struggle to fill gaps for the poor and under-resourced whether providing care in emergency departments, critical care settings, school nurse offices, or community health centers. Justice is a core bioethical principle, and treating individuals with dignity, respect, and what is due or owed to them is a professional and moral obligation; yet frontline providers are constantly navigating the societal morass and gaps in resources for their patients and families. Lack of resources for mental health care, housing, and services for the homeless are examples where nurses, social workers, physicians, and other care providers can find themselves in ethically compromising situations when resources are limited or not available. Unfortunately, compromising care and responding to crises because there is a lack of health services results in clinicians' moral distress. Explaining to a patient that their insurance does not cover a needed medication or treatment or responding to pressures of competing obligations by creating a workaround to stave off an immediately foreseeable problem is ethically problematic. Today clinicians face competing pressures—to deliver value, be efficient, do more with less, make it work, get the job done, and be a team player—often compelling nurses and other clinicians to improvise short-term solutions to relieve these pressures [78]. This often places them in difficult positions and may promote moral distress.

Conclusion

Identifying and understanding the many root causes of moral distress helps to target specific areas for interventions that might diminish the impact of moral distress within the clinical environment. Moral distress may be experienced wherever and whenever clinicians are obligated to provide health care. As an Emergency Department (ED) nurse noted: "continuing aggressive care for a 90 year old woman with terrible pain was the worst experience in my 7.5 years of practice in the ED." Although the root causes may vary, and new sources will likely be recognized, the end result of moral distress is profound; indeed, it has the power to negatively impact all institutional life and the quality within and beyond those walls—individual nurses, physicians, and other care providers, interprofessional and team practice and goals, and ultimately the care we provide to our patients and their families.

References

1. Jameton A. Nursing practice: the ethical issues. Prentice Hall: Englewood Cliffs; 1984.
2. Moss M, Good VS, Gozal D, Kleinpell R, Sessler CNA. Critical care societies collaborative statement: burnout syndrome in critical care health-care professionals. a call for action. Am J Respir Crit Care Med. 2016;194(1):106–13.
3. Rushton CH. Moral resilience: a capacity for navigating moral distress in critical care. AACN Advanced Critical Care. 2016;27(1):111–9.
4. Weigand DL, Funk M. Consequences of clinical situations that cause critical care nurses to experience moral distress. Nurs Ethics. 2012;19(4):479–87.
5. Epstein EG, Hamric AB. Moral distress, moral residue, and the crescendo effect. J Clin Ethics. 2009;20(4):330–42.
6. Allen R, Judkins-Cohn T, de Velasco R, Forges E, Lee R, Clark L, et al. Moral distress among healthcare professionals at a health system. JONA's Healthcare Law Ethics Regul. 2013;15(3):111–8. quiz 9–20
7. Henrich NJ, Dodek PM, Alden L, Keenan SP, Reynolds S, Rodney P. Causes of moral distress in the intensive care unit: a qualitative study. J Crit Care. 2016;35:57–62.
8. Houston S, Casanova MA, Leveille M, Schmidt KL, Barnes SA, Trungale KR, et al. The intensity and frequency of moral distress among different healthcare disciplines. J Clin Ethics. 2013;24(2):98–112.
9. Whitehead PB, Herbertson RK, Hamric AB, Epstein EG, Fisher JM. Moral distress among healthcare professionals: report of an institution-wide survey. J Nurs Scholarsh. 2015;47(2):117–25.
10. Hamric AB. Empirical research on moral distress: issues, challenges, and opportunities. HEC Forum. 2012;24:39–49.
11. Hanna DR. Moral distress: the state of the science. Res Theory Nurs Pract. 2004;18(1):73–93.
12. Campbell SM, Ulrich CM, Grady C. A broader understanding of moral distress. AJOB. 2016;16(12):2–9.
13. Varcoe C, Pauly B, Webster G, Storch J. Moral distress: tensions as springboards for action. HEC Forum. 2012;24(1):51–62.
14. Cribb A. Integrity at work: managing routine moral stress in professional roles. Nurs Philos. 2011;12(2):119–27.
15. Berlinger N. Are workarounds ethical? Managing moral problems in health care systems. New York: Oxford University Press; 2016.
16. Epstein EG, Hurst AR. Looking at the positive side of moral distress: why it's a problem. J Clin Ethics. 2017;28(1):37–41.
17. Carse A, Rushton CH. Harnessing the promise of moral distress: a call for re-orientation. J Clin Ethics. 2017;28(1):15–29.
18. Elpern EH, Covert B, Kleinpell R. Moral distress of staff nurses in a medical intensive care unit. Am J Crit Care. 2005;14(6):523–30.
19. Gutierrez KM. Critical care nurses' perceptions of and responses to moral distress. Dimens Crit Care Nurs. 2005;24(5):229–41.
20. Hefferman P, Heilig S. Giving "moral distress" a voice: ethical concerns among neonatal intensive care unit personnel. Camb Q Healthc Ethics. 1999;8(2):173–8.
21. Rushton CH. Moral resilience: a capacity for navigating moral distress in critical care. Adv Crit Care. 2016;27(1):111–9.
22. Oh Y, Gastmans C. Moral distress experienced by nurses: a quantitative literature review. Nurs Ethics. 2015;22(1):15–31.
23. McAndrew NS, Leske J, Schroeter K. Moral distress in critical care nursing: the state of the science. Nurs Ethics. 2016. https://doi.org/10.1177/0969733016664975.
24. Rice EM, Rady MY, Hamrick A, Verheijde JL, Pendergast DK. Determinants of moral distress in medical and surgical nurses at an adult acute tertiary care hospital. J Nurs Manag. 2008;16(3):360–73.

25. Powell SB, Engelke MK, Swanson MS. Moral distress among school nurses. J Sch Nurs. 2017;2017:1059840517704965.
26. Robinson R, Stinson CK. Moral distress: a qualitative study of emergency nurses. Dimens Crit Care Nurs. 2016;35(4):235–40.
27. Wolf L, Perhats C, Delao A, Moon MD, Clark PR. Zavotsky. "It's a burden you carry": describing moral distress in emergency nursing. J Emerg Nurs. 2016;42(1):37–46.
28. Hamilton Houghtaling DL. Moral distress: an invisible challenge for trauma nurses. J Trauma Nurs. 2012;19(4):232–7. quiz p 8–9
29. Sasso L, Bagnasco A, Bianchi M, Bressan V, Carnevale F. Moral distress in undergraduate nursing students: a systematic review. Nurs Ethics. 2016;23(5):523–34.
30. Loomis K, Carpenter R, Miller B. Moral distress in the third year of medical school: a descriptive review of student case reflections. Am J Surg. 2009;197:107–12.
31. Dzeng E, Colaianni A, Roland M, Levine D, Kelly MP, Barclay S, et al. Moral distress amongst american physician trainees regarding futile treatments at the end of life: a qualitative study. J Gen Intern Med. 2016;31(1):93–9.
32. Sajjadi S, Norena M, Wong H, Dodek P. Moral distress and burnout in internal medicine residents. Can Med Educ J. 2017;8(1):e36–43.
33. McCarthy J, Gastmans C. Moral distress: a review of the argument-based nursing ethics literature. Nurs Ethics. 2015;22(1):131–52.
34. Austin W. The ethics of everyday practice: healthcare environments as moral communities. Adv Nurs Sci. 2007;30(1):81–8.
35. Hamric AB. Moral distress in everyday ethics. Nurs Outlook. 2000;48(5):199–201.
36. Komesaroff P. Experiments in love and death – medicine, postmodernism, microethics and the body. Austin: River Grove Books; 1995.
37. Truog RD, Brown SD, Browning D, Hundert EM, Rider EA, Bell SK, et al. Microethics: the ethics of everyday clinical practice. Hast Cent Rep. 2015;45(1):11–7.
38. Zizzo N, Bell E, Racine E. What is everyday ethics? A review and a proposal for an integrative concept. J Clin Ethics. 2016;27(2):117–28.
39. Benner P, Tanner C, Chesla C. Expertise in nursing practice – caring, clinical judgment, and ethics. 2nd ed. New York: Springer Publishing Company; 2009. p. 58–9.
40. Tulsky JA, Beach MC, Butow PN, Hickman SE, Mack JW, Morrison RS, et al. A research agenda for communication between health care professionals and patients living with serious illness. JAMA Intern Med. 2017;177(9):1361–6.
41. Schulz R. In: Schulz R, Eden J, editors. Families caring for an aging America. Washington DC: National Academies Press; 2016.
42. Gillick MR. The critical role of caregivers in achieving patient-centered care. JAMA. 2013;310(6):575–6.
43. Browning AM. CNE article: moral distress and psychological empowerment in critical care nurses caring for adults at end of life. Am J Crit Care. 2013;22(2):143–51.
44. Ferrell BR. Understanding the moral distress of nurses witnessing medically futile care. Oncol Nurs Forum. 2006;33(5):922–30.
45. Hamric AB, Blackhall LJ. Nurse-physician perspectives on the care of dying patients in intensive care units: collaboration, moral distress, and ethical climate. Crit Care Med. 2007;35(2):422–9.
46. IOM. Dying in America: Improving quality and honoring individual prefererences near end of life. Washington, DC: National Academies Press; 2015.
47. Fairman J, Lynaugh JE. Critical care nursing a history. Philadelphia: University of Pennsylvania Press; 1998. p. 6.
48. Bosslet GT, Pope TM, Rubenfeld GD, Lo B, Truog RD, Rushton CH, et al. An official ATS/AACN/ACCP/ESICM/SCCM policy statement: responding to requests for potentially inappropriate treatments in intensive care units. Am J Respir Crit Care Med. 2015;191(11):1318–30.
49. Bruce CR, Miller SM, Zimmerman JL. A qualitative study exploring moral distress in the ICU team: the importance of unit functionality and intrateam dynamics. Crit Care Med. 2015;43(4):823–31.
50. Traudt T, Liaschenko J, Peden-McAlpine C. Moral agency, moral imagination, and moral community: antidotes to moral distress. J Clin Ethics. 2016;27(3):201–13.

51. Olsen DP, Chan R, Lehto R. Ethical nursing care when the terminally ill patient seeks death. Am J Nurs. 2017;117(7):50–5.
52. Fiester A. Teaching nonauthoritarian clinical ethics: using an inventory of bioethical positions. Hast Cent Rep. 2015;45(2):20–6.
53. Stein LI. The doctor-nurse game. Arch Gen Psychiatry. 1967;16(6):699–703.
54. Interprofessional Education Collaborative. Core competencies for interprofessional collaborative practice: 2016 update. Washington, DC: Interprofessional Education Collaborative; 2016.
55. Hamric AB. Reflections on being in the middle. Nurs Outlook. 2001;49(6):254–7.
56. Murphy P. Clinical ethics: must nurses be forever in the middle? Bioethics Forum. 1993;9(4):3–4.
57. Varcoe C, Doane G, Pauly B, Rodney P, Storch JL, Mahoney K, et al. Ethical practice in nursing: working the in-betweens. J Adv Nurs. 2004;45(3):316–25.
58. Berkowitz KA, Katz AL, Powderly KE, Spike JP. Quality assessment of the ethics consultation service at the organizational level: accrediting ethics consultation services. Am J Bioeth. 2016;16(3):42–4.
59. Hamric AB, Wocial LD. Institutional ethics resources: creating moral spaces. Hast Cent Rep. 2016;46(Suppl 1):S22–7.
60. Grady C, Danis M, Soeken KL, O"Donnell P, Taylor C, Farrar A, et al. Does ethics education influence the moral action of practicing nurses and social workers. Am J Bioeth. 2008;8(4):4–11.
61. Robinson EM, Lee SM, Zollfrank A, Jurchak M, Frost D, Grace P. Enhancing moral agency: clinical ethics residency for nurses. Hast Cent Rep. 2014;44(5):12–20.
62. Verkerk M, Lindemann H, Maeckelberghe E, Feenstra E, Hartoungh R, De Bree M. Enhancing reflection: an interpersonal exercise in ethics education. Hast Cent Rep. 2004;34(6):31–8.
63. Fowler MMDM. Guide to the Code of Ethics for Nurses with Interpretative Statements, 2nd Ed. Silver Spring, MD: American Nurses Association. 2015.
64. Institute of Medicine (IOM). Crossing the quality chasm: A new health system for the 21st century. Washington DC: National Academies Press; 2001.
65. Cronenwett L, Sherwood G, Barnsteiner J, Disch J, Johnson J, Mitchell P, et al. Quality and safety education for nurses. Nurs Outlook. 2007;55(3):122–31.
66. Lines LM, Lepore M, Wiener JM. Patient-centered, person-centered, and person-directed care: they are not the same. Med Care. 2015;53(7):561–3.
67. Scheunemann LP, Arnold RM, White DB. The facilitated values history: helping surrogates make authentic decisions for incapacitated patients with advanced illness. Am J Respir Crit Care Med. 2012;186(6):480–6.
68. American Nurses Association. Position Statement: Incivility, Bullying and Workplace Violence. 2015.
69. Leape LL, Shore MF, Dienstag JL, Mayer RJ, Edgman-Levitan S, Meyer GS, et al. Perspective: a culture of respect, part 1: the nature and causes of disrespectful behavior by physicians. Acad Med. 2012;87(7):845–52.
70. The Joint Commission. Behaviors that undermine a culture of safety. Sentinel Event Alert; 2008.
71. Berlinger N. When policy produces moral distress: reclaiming conscience. Hast Cent Rep. 2016;46(2):32–4.
72. Corley MC, Minick P, Elswick RK, Jacobs M. Nurse moral distress and ethical work environment. Nurs Ethics. 2005;12(4):381–90.
73. Rushton CH, Caldwell M, Kurtz MCE. Moral distress: a catalyst in building moral resilience. Am J Nurs. 2016;116(7):40–9.
74. Hamric AB, Epstein EGA. Health system-wide moral distress consultation service: development and evaluation. HEC Forum. 2017;29(2):127–43.
75. Sabin JE. Using moral distress for organizational improvement. J Clin Ethics. 2017;28(1):33–6.
76. Shanafelt TD, Noseworthy JH. Executive leadership and physician well-being: nine organizational strategies to promote engagement and reduce burnout. Mayo Clin Proc. 2017;92(1):129–46.
77. Mason DJ. Promoting the health of families and communities: a moral imperative. Hast Cent Rep. 2016;46(Suppl 1):S48–51.
78. Berlinger N. Workarounds are routinely used by nurses-but are they ethical? Am J Nurs. 2017;117(10):53–5.

Building Compassionate Work Environments: The Concept of and Measurement of Ethical Climate

Linda L. Olson

6.1 Building Compassionate Work Environments: The Concept and Measurement of Ethical Climate

Ethical climate can be defined as the organizational conditions and practices that influence the ways in which ethical issues and concerns are identified, discussed, and decided [1]. It is a type of organizational climate, which derives from the focus on ethics and ethical practices in an organization. This concept is derived from Schneider's concept of types of organizational climates, which states that organizations have not one but many types of climates [2]. The type of climate is based on one's area of strategic interest; thus, an interest in perceptions of the way ethical issues and concerns are handled in an organization is referred to as ethical climate. The purpose of this paper is to define the concept of and a measure of ethical climate as a component of ethical work environments in healthcare organizations. In addition, it is important to differentiate between the concepts of ethical climate and ethical culture and to use the nursing profession as an example in highlighting how the nursing profession's Code of Ethics for Nurses with Interpretive Statements [3] provides support for nurses involved in difficult ethical issues. Research that used the Hospital Ethical Climate Survey (HECS) as a measure of perceptions of the mechanisms in place for supporting ethical practice and of the influence of the workplace on ethical practice will be highlighted. The HECS has been used in research with nurses as well as other members of the healthcare team, all who contribute to and are influenced by the ethical climate of their work setting. The Code of Ethics provides a framework within which the concept of ethical climate can be understood.

L.L. Olson
Nursing Regulation, National Council of State Boards of Nursing (NCSBN),
Chicago, IL, USA

Member of American Nurses Association Ethics Advisory Board,
Silver Spring, MD, USA
e-mail: lolson@ncsbn.org; lalolson@gmail.com

© Springer International Publishing AG 2018   95
C.M. Ulrich, C. Grady (eds.), *Moral Distress in the Health Professions*,
https://doi.org/10.1007/978-3-319-64626-8_6

Ethics occurs in the context of relationships; it is relationship-based. The key relationships are those with whom healthcare providers interact in their work environment. These include relationships with colleagues and other co-workers, with physicians, nurses, managers, other healthcare providers, patients and families, as well as with the overall organization. The concept of ethics also occurs within the context of healthcare providers' personal and professional values and the core values of their profession [4]. Provision 6 of the ANA Code of Ethics for Nurses with Interpretive Statements [3], for example, states that the nurse, through individual and collective effort, establishes, maintains, and improves the ethical environment of the work setting and conditions of employment that are conducive to safe, quality healthcare (p. 23). Since the work environment influences behavior, managers and leaders in the organization are responsible for creating an ethical work environment and for establishing working conditions that promote safe practice and collaborative interprofessional relationships. Leaders provide the resources to implement the structures and programs that support ethics and ethical practice. Through their behavior as role models and in recognition that actions speak louder than words, managers enact their role as ethical leaders.

6.2   Measure of Ethical Climate: The Hospital Ethical Climate Survey (HECS, [1])

The Hospital Ethical Climate Survey (HECS) is a 26-item survey, in which the items or variables are categorized into five key relationships that nurses and all healthcare professionals have within the work environment: colleagues, managers, physicians, patients and families, and the hospital (or relevant healthcare organization). The items represent workplace conditions that, when present, facilitate healthcare professionals and others to engage in ethical reflection and decision-making about difficult issues that arise in patient care or in relations with others. In addition to Schneider's [2] work on types of organizational climates, the framework that guided instrument development also included Brown's [5] conditions for ethical reflection in organizations. Brown purports that, for ethical reflection to occur, the conditions of power, trust, inclusion, role flexibility, and inquiry must be present. In the next section, these conditions will be defined and exemplified using the Code of Ethics for Nurses with Interpretive Statements (2015).

6.3   Conditions for Ethical Reflection in Organizations

6.3.1   Power and Trust

Healthcare employees have the right to receive relevant information and to be free to say what needs to be said about an issue. This is the condition of power, which is necessary for ethical reflection to occur among organizational participants [5]. In addition, they must be able to trust one another in order to be free to disagree and engage in discussion in order to increase their understanding of issues. Provision 1 of the ANA Code of Ethics is an example of this because it addresses the concept of respect, which includes the nurse's role in establishing relationships of trust with patients, colleagues, and others. The condition of trust facilitates the healthcare team's understanding of issues and confidence that they can express their viewpoints without fear of retaliation or of their disagreements being used against them [5]. In building a compassionate work environment, and one in which all members of the healthcare team participate in ethical reflection and decision-making, it is important to create conditions where each participant feels they are listened to and respected. Building trust among all members of the healthcare team is an essential component to creating an ethical climate, one in which all those who have an interest in and involvement in an issue can feel free to participate and express their viewpoints.

6.3.2 Inclusion

The condition of inclusion is met when individuals and groups who have a role or interest in the issue and the outcome of a decision are involved in the conversation and decision-making processes. These include nurses, physicians, other healthcare colleagues, as well as patients and families and other appropriate stakeholders. Inclusion is discussed in Provision 4 of the ANA Code of Ethics, where it states that "the nurse has authority, accountability, and responsibility for nursing practice; makes decisions; and takes action consistent with the obligation to promote health and to provide optimal care" (p. 15). Section 4.3, which discusses responsibility for nursing judgments, decisions, and actions, states "nurses must bring forward difficult issues related to patient care and/or institutional constraints upon ethical practice for discussion and review. The nurse acts to promote inclusion of appropriate individuals in all ethical deliberations" (p. 16). Section 1.4 of the Code emphasizes that the role of nurses is to include patients or surrogate decision-makers in discussions and to support patient autonomy and decision-making processes (p. 3). Just as nurses expect and desire to be included in decisions affecting their patients and families, all members of the healthcare team who have an interest in and role with identified patients and families have a right to be included in discussions and decision-making.

6.3.3 Role Flexibility and Inquiry

The condition of role flexibility implies that participants in decision-making about ethical issues are allowed to change their views or to take different positions. Similarly, the condition of inquiry is present when participants are encouraged to ask questions to gain the needed information for informed decision-making. An example of this is Provision 5 of The ANA Code of Ethics, which addresses the concept that "the nurse owes the same duties to self as to others" (p. 19). Section 5.3, which discusses preservation of wholeness of character, states that "sound ethical decision-making requires the respectful and open exchange of views among all those with relevant interests," including the nurse's responsibility "to express moral perspectives" that apply to the issue, "whether or not those perspectives are shared by others" (p. 20). This provision also addresses the responsibility of nurses who experience moral distress related to institutional or professional practices to report their concerns to an "appropriate authority or committee" (p. 21).

6.3.4 Ethical Climate or Ethical Culture

The terms ethical climate and ethical culture are often used interchangeably; however they are distinct and separate concepts. Ethical culture can be viewed as the way ethical issues and situations creating ethical concerns are handled in an organization. Hospital ethics committees or ethics consultants are mechanisms in place within a facility that comprise the way ethical issues should be handled. Ethical climate constitutes employee perceptions of these organizational practices. If healthcare providers perceive that an ethics committee is either not accessible to them or that they need explicit permission from others to access it, it may not be perceived as a helpful mechanism for addressing ethical issues or allowing for ethical reflection. Therefore, ethical climate constitutes employee perceptions of how decisions having ethical content are discussed and resolved and of the support offered within the workplace for engaging in ethical reflection and problem-solving.

Provision 6 of the ANA Code of Ethics, for example, addresses the need for ethical practice environments, stating that "the nurse, through individual and collective effort, establishes, maintains, and improves the ethical environment of the work setting and conditions of employment that are conducive to safe, quality health care" (p. 23). This provision upholds the importance of an ethical environment in which the ethical practice of nurses and others is essential to meeting the preferences and goals of patients and families. Nurses in all roles and settings are responsible for contributing to an environment in which colleagues and peers interact in a respectful manner that facilitates open discussion of ethical issues. Those issues can involve interactions with patients and families or decisions related to nursing practice and working conditions (p. 24). The particular responsibility of healthcare executives is to assure the fair and just treatment of all employees as well as to provide mechanisms for nurses and others to address concerns about the healthcare environment. Similarly, all healthcare employees contribute to the creation and continual improvement of the ethical environment of their workplaces.

6.4 Research

By using a measure of ethical climate, researchers can study the influence of the work environment on nurses' ethical practice [1]. Such a measure has also been used in research with other healthcare professionals. Researchers have studied relationships among ethical climate and moral distress, nurse turnover or nurses' intent to leave their position or the profession [6], job satisfaction [7], collaboration [8], moral sensitivity and work-related moral stress [9], nurse competence [10], and medical error [11]. The research has been conducted with nurses and others in hospitals, long-term care [12, 13], mental health settings [9], and others.

6.4.1 The Relationship Between Ethical Climate and Moral Distress

A number of nurse researchers have studied the relationship between ethical climate and moral distress [8, 14–16]. Moral distress occurs when the ethically correct action is known; however, situations in the workplace prevent nurses from carrying out the action they believe is appropriate. Fogel [14], who also studied turnover intentions in critical care nurses in two hospitals, found that the more positive the perceptions of ethical climate, the lower the likelihood that nurses' scores on intent to turnover were high. Ethical climate factors, such as relationships with peers and managers, moderated the effect of moral distress on nurses' intent to leave their positions [14]. Thus, the way in which nurses perceive the ethical climate of their work environment is inversely related to moral distress. The higher and more positive the perceptions of ethical climate, the less likely that nurses experience levels of moral distress that lead to the likelihood of leaving their positions. In their study of 374 nurses in acute care hospitals in British Columbia, Pauly et al. [15] also found that higher scores on ethical climate resulted in less intense levels of moral distress as measured by Corley et al.'s [17] Revised Moral Distress Scale. In addition, a study on 249 nurses in 2 hospitals in Sweden found that the more positive the perceptions of ethical climate, the less frequent were nurses' reporting of morally distressing situations [16].

6.5 Ethical Climate and Turnover Intention and Job Satisfaction

Positive perceptions of ethical climate are associated with lower turnover intentions and nurses reporting higher intention to stay in their positions [6, 11, 18]. Perceptions of ethical climate can be managed, changed, and improved. For example, managers can improve the ethical climate by providing support for nurses and other healthcare professionals to actively participate in decision-making about patient care with physicians. Research has also demonstrated that the more positive the perceptions of ethical climate, the higher the reported level of job satisfaction by nurses [7].

In a study of 1826 nurses in 33 public hospitals in South Korea, Hwang and Park [11] found that nurses with a more positive perception of the ethical climate dimension related to "patients" were less likely to report making medical errors. Korean nurses rated the ethical climate in their hospital an average of 3.6 out of 5. Of the five dimensions measured by the Hospital Ethical Climate Scale, the score on the "physician" dimension was lowest (3.0) in this study, indicating "…the need for increased collaboration between nurses and physicians and for promoting mutual support, respect and shared decision making regarding patient care" [11]. When studies find lower scores on the dimension with physicians in the HECS, improving nurse-physician relationships can then be identified and implemented. In this way, the HECS can be used as a tool to identify areas in which the professional work environment can be managed, changed, and improved. Future research is needed to better understand how interprofessional teams enhance the ethical climate of the workplace and improve quality of care delivery.

6.6   Summary and Conclusion

Nurses and all members of the healthcare team face increasing demands in a workplace environment where patient needs and their corresponding care are complex and challenging. Whether referred to as a healthy work environment, a compassionate work environment, or an ethical work environment, it is important to create one that contributes to positive patient and healthcare employee outcomes. Positive perceptions of ethical climate can mitigate the impact of moral distress associated with difficult and complex ethical issues in the healthcare workplace.

References

1. Olson LL. Hospital nurses' perceptions of the ethical climate of their work setting. J Nurs Scholarsh. 1998;30(4):345–9.
2. Schneider B. Organizational climate and culture. San Francisco: Jossey-Bass; 1990.
3. American Nurses Association. Code of Ethics for Nurses with Interpretive Statements. Silver Spring, MD: ANA; 2015.
4. Olson LL. Standard 7: ethics. In: White KM, O'Sullivan A, editors. The essential guide to nursing practice: applying ANA's scope and standards in practice and education. Silver Spring: American Nurses Association; 2012.
5. Brown MT. Working ethics: strategies for decision making and organizational responsibility. Oakland: Regent Press; 2000.
6. Hart S. Hospital ethical climates and registered nurses' turnover intentions. J Nurs Scholarsh. 2005;37(2):173–7.
7. Joolaee S, Jalili HR, Rafii F, Hajibabaee F, Haghani H. The relationship between ethical climate at work and job satisfaction among nurses in Tehran. Indian J Med Ethics. 2013;10(4):238–42.
8. Hamric AB, Blackhall LJ. Nurse-physician perspectives on the care of dying patients in intensive care units: collaboration, moral distress, and ethical climate. Crit Care Med. 2007;35(2):422–9.
9. Lützén K, Blom T, Ewalds-Kvist B, Winch S. Moral stress, moral climate and moral sensitivity among psychiatric professionals. Nurs Ethics. 2010;17(2):213–24.
10. Numminen O, Leino-Kilpi H, Isoaho H, Keretoja R. Ethical climate and nurse competence-newly graduated nurses' perceptions. Nurs Ethics. 2015;22(8):845–959.
11. Hwang J, Park H. Nurses' perception of ethical climate, medical error experience and intent-to-leave. Nurs Ethics. 2014;21(1):28–42.
12. Suhonen R, Stolt M, Gustafsson ML, Katajisto J, Charalambous A. The associations among the ethical climate, the professional practice environment and individualized care in care settings for older people. J Adv Nurs. 2014;70(6):1356–68.

13. Suhonen R, Stolt M, Jouko K, Charalambous A, Olson L. Validation of the hospital ethical climate survey for older people care. Nurs Ethics. 2015;22(5):517–32.
14. Fogel, K.M. The relationships of moral distress, ethical climate, and intent to turnover among critical care nurses. PhD doctoral dissertation, Loyola University Chicago; 2007. ISBN: 9780549227717.
15. Pauly B, Varcoe C, Storch J, Newton L. Registered Nurses' perceptions of moral distress and ethical climate. Nurs Ethics. 2009;16(5):561–73.
16. Silén M, Svantesson M, Kjellström S, Sidenvalli B, Christensson L. Moral distress and ethical climate in a Swedish nursing context: perceptions and instrument usability. J Clin Nurs. 2011;20(23/24):3483–93.
17. Corley MC, Elswick RK, Gorman M, Clor T. Development and evaluation of a moral distress scale. J Adv Nurs. 2001;33:250–6.
18. Ulrich C, O'Donnell P, Taylor C, Farrar A, Danis M, Grady C. Ethical climate, ethics stress, and the job satisfaction of nurses and social workers in the United States. Soc Sci Med. 2007;65(8):1707–19.

Carol L. Pavlish, Ellen M. Robinson, Katherine Brown-Saltzman, and Joan Henriksen

7.1 Introduction

Moral distress appears to arise in situations where individuals believe they have violated their core moral obligations, compromising their moral integrity [1]. Healthcare clinicians' professional and moral obligations reside in beliefs about virtuous character, professional identity, and knowledge of one's professional code of ethics. Standards of practice such as the Quality and Safety Education for Nurses (QSEN) initiative, funded by the Robert Wood Johnson Foundation, further specify professional and moral obligations [2]. The QSEN movement grew out of the Institute of Medicine's report on *Health Professions Education* which specified health professionals' identity as providing patient-centered care as part of interdisciplinary teams emphasizing safety, evidence-based practice, quality improvement, and informatics [3]. Fidelity to these intersecting obligations, however, can be quite challenging. When patient care violations occur, moral distress and/or moral disengagement accompanied by safety and quality concerns often result [4] which, in turn, can set up a downward cycle that further compromises patients' well-being and clinicians' moral integrity.

C.L. Pavlish (✉)
University of California Los Angeles, School of Nursing, Los Angeles, CA, USA
e-mail: cpavlish@sonnet.ucla.edu

E.M. Robinson
Massachusetts General Hospital, Boston, MA, USA

K. Brown-Saltzman
University of California Los Angeles, Health Ethics Center, Los Angeles, CA, USA

J. Henriksen
Mayo Clinic, Rochester, MN, USA

© Springer International Publishing AG 2018
C.M. Ulrich, C. Grady (eds.), *Moral Distress in the Health Professions*,
https://doi.org/10.1007/978-3-319-64626-8_7

103

Recognizing moral distress among critical care nurses, the American Association of Critical-Care Nurses issued a position statement on moral distress in 2008 [5]. More recently, four critical care societies issued a joint call to action to implement strategies that mitigate the development of moral distress and clinician burnout [6]. Furthermore, quality experts propose that improving the care and work experiences for healthcare professionals is a prerequisite condition for the "Triple Aim" of improving the patient experience, improving population health, and reducing costs [7]. Bodenheimer and Simsky [8] describe the Quadruple Aim that includes this fourth domain of attention to the work life and well-being of healthcare providers and teams as an essential aspect of health system performance. In response, the National Academy of Medicine initiated an Action Collaborative on Clinician Well-Being and Resilience [9]. Part of a strategy to address the fourth aim is a robust research agenda not only on moral distress but also on all system processes, organizational practices, and individual attributes that promote professional identity and moral integrity.

In this chapter, we draw on a socioecological perspective to suggest a research agenda for moral distress. Lutzen and Kvist [10] argued that a conceptual framework rather than a consensus definition of moral distress advances our understanding of the phenomenon. Others have noted that relationships between major moral distress concepts such as moral agency and ethical climate merit further study [11]. To address these concerns, we present a framework that conceptualizes moral distress as one component in a spectrum of experiences that occur during ethically challenging situations. The crescendo effect described by Epstein and Hamric [12] suggests that experiences over time influence the moral distress experience. However, the process of developing moral distress and understanding its impact requires more study [13]. Our aim is to consider moral distress within the constructs of time and space and a larger context of moral obligations, agency, and integrity.

The proposed framework also offers direction for addressing moral distress through research on prevention, risk reduction, mitigation, and treatment. Scholars note that moral distress interventions have not been adequately developed or studied [13, 14]. Our goal is to discover novel time-points for action that interrupts the trajectory of repetitive moral distress and accumulating moral residue. Whether gradual or sudden, moral distress unfolds over time; therefore, a variety of evidence-based strategies, resources, and supports should be readily available. We believe that widening the research lens on moral distress will offer important opportunities to create change within and between all levels of the healthcare system—among individual patients who require care, surrogates and families who need support, clinicians who provide care, systems that arrange and support care, and the public who have a right to quality care. We also offer examples of foundational moral distress research from which researchers and quality improvement specialists can design meaningful and productive research questions and improvement strategies. Finally, we suggest a monitoring device for assessing progress on moral distress science in a systematic, coordinated, and collaborative effort.

Programs and activities that assist clinicians to practice from a position of moral agency and integrity are critical for high-quality ethical patient care, as well as to decrease burnout and retain clinicians in the professional workforce [6, 15]. Mechanisms to support clinicians are appearing in healthcare organizations, but their impact on decreasing moral distress needs further study.

7.2   A Theoretical Perspective for Researching Moral Distress

To frame the research agenda, we adopt a socioecological perspective, which assumes that individuals are active agents who not only shape but are also influenced by their environments. As such, individuals with varied cultural histories, skills, dispositions, values, and identities interact within the opportunities, resources, and constraints of their social contexts [16]. Dimensions of individual integrity and well-being are reciprocally related and linked to diverse and evolving conditions in the sociopolitical environment including the quality of human interactions. Congruity of values and actions within persons, the quality of interactions between persons, and the nature of conditions surrounding persons all influence persons' ability to preserve their moral integrity. Because ethical issues occur within multilayered and interdependent "contexts of practice" with interacting influencing factors [17] (p. 323), a socioecological perspective seems particularly relevant for proposing a research agenda on moral distress and its multilayered, situational conditions. An ecological perspective widens the research lens by calling attention to dynamic, interdependent, multidimensional, multi-level, and interactional views of phenomena—in this case, moral integrity and moral distress.

Particular constructs pertaining to time and space are suggested with a socioecological approach [18]. *First, embodiment* and its *pathways* refer to the ways that we biologically incorporate the social world in which we live including societal arrangements of power and position with resultant opportunities and constraints. Moral integrity and distress are human experiences that occur within social structures and result from relationships and events that evolve into pathways over time and across interpersonal space. *Second, internal and external factors* also influence pathway development over time and across space. This pathway reflects a *dynamic, cumulative interaction* between exposure, susceptibility, adaptation, and resistance to influencing factors. To illustrate this point, consider the everyday moral work of clinicians as a pathway that constantly evolves while influenced by the interplay of factors such as patient condition and needs, team dynamics, conflicting values, resource availability and allocation, risk management, administrative practices, communication opportunities, and family distress and demands. The degree of exposure, susceptibility, resistance, and adaptation to these interweaving and multi-level factors form a pathway that is simultaneously experienced by and can be influenced by healthcare providers as they work with others in the moral spaces and time of clinical practice. For example, clinicians who are "exposed" to ethically difficult situations and are unable to voice their perspective or problem solve what they perceive as threats to their moral integrity may become more "susceptible" to moral

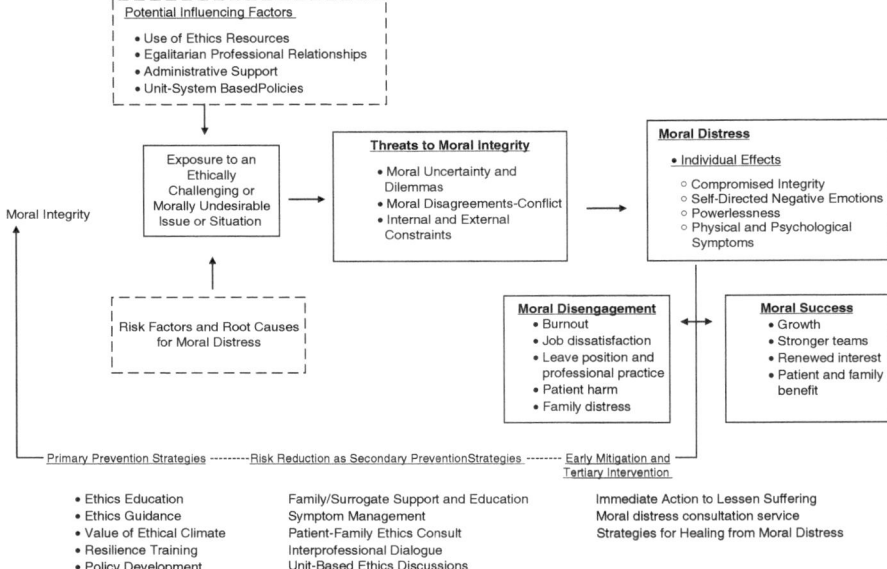

Fig. 7.1 Framework for moral distress research

distress and disengagement. *Finally*, the concepts of *accountability and agency* pertain to pathway development as individuals interact within the context of institutions, professional identities and responsibilities, and systems of care.

Taken together, ecological constructs such as embodiment, pathway, moderating factors, cumulative interactions, accountability, and agency expand our avenues for research methods such as factor analysis, correlational analyses, structural equation modeling (e.g., see Rathert [19]), and intervention studies that manipulate internal and external factors to construct improved pathways. Additional methods for measuring one's moral experience across time, space, and relationships can also be developed and tested. As noted recently by Elizabeth Peter, "It is not that moral distress is no longer relevant, but we need to expand our understanding through additional concepts that help us understand the ethics of nursing work with its frequent proximity to patients and its political positioning in a variety of settings" [20] (p. 3). The spatial-temporal concepts in the socioecological perspective potentially broaden our study of moral distress without diluting or devaluing the concept's significance to clinical practice.

7.3   A Socioecological Research Agenda for Moral Integrity and Moral Distress

Figure 7.1 depicts moral distress as a downstream concept that evolves from being exposed to and immersed in situations that threaten to compromise moral integrity. Internal and external factors influence the quality of interactions and determine the pathway's course, which also depends on exposure, susceptibility, adaptation, resistance, agency, and accountability over time and space within a specific context. Along this complex pathway, experiences such as moral uncertainty, moral dilemma, moral disagreement, mild or delayed distress, and moral association could lead to moral distress (see Chap. 4 by Campbell). Theoretically, if the number or intensity of intersecting factors pushing clinicians toward moral distress is greater than the experiences that tend to preserve moral integrity, the cumulative effect begins to shift the pathway toward moral distress.

7.3.1 Defining and Measuring Moral Distress

Scholars have proposed different definitions of moral distress [11, 21–24] (see Chap. 2). Many similarities exist among these various definitions. The proposed research framework in Fig. 7.1 attempts to honor Jameton's classic moral distress definition [23] while also embracing tenets from other prominent and noted moral distress scholars and researchers. As we learn more about the pathway of becoming morally distressed, we will also learn more about moral distress as an outcome that occurs in healthcare situations.

Measuring moral distress is a dynamic process. Items in the classic Moral Distress Scale Revised (MDS-R) continue to be assessed for relevance, validity, and reliability [25]. Other measures have been suggested such as the Moral Distress Thermometer [26]. Instruments that measure other components of clinicians' moral work such as moral integrity and its influencing factors moral responsibilities, ethics self-efficacy, and interprofessional ethics practices need to be developed to fully realize the context in which moral distress unfolds and is experienced. Continuing to develop measurement metrics that are conceptually sound yet practically low-burden can facilitate the development of strategies to address moral distress. Organizational leaders need concise, action-guiding metrics that can be assessed longitudinally in order to incorporate them into organizational planning, culture, and human resource strategies. For example, moral distress assessment questions could be incorporated into standard periodic all-staff surveys so institutions could trend data over time and develop appropriate action responses.

This chapter is geared toward a research agenda; however, quality improvement (QI) projects can also contribute important lessons for more formal research on moral integrity and distress. QI methods have the salient feature of focusing on a *local* problem, which is a very important characteristic of moral distress—its sources are often dependent on the *local culture and ethical climate*. Effective solutions can also emerge from local changes. QI methods also tend to be more flexible and responsive to real-time problems that arise. Data that emerges from soundly designed and implemented QI projects can be important information for innovative intervention development that can be tested more rigorously in subsequent studies. The following sections provide reflections and specific ideas for advancing the science related to moral integrity and distress.

7.3.2 Exploring Moral Integrity

To date, moral distress studies and systematic reviews have primarily examined the experiences, levels, influencing factors, and consequences of clinicians' moral distress, especially for nurses in various settings [6, 13, 19, 20, 27–32]. In other words, upstream concepts such as moral integrity and what preserves or disrupts integrity during ethically challenging situations remain relatively unexamined. A few recent initiatives have explored upstream factors. For example, Traudt and colleagues examined the experiences of critical care nurses who managed to successfully "avoid or navigate through" moral distress [33] (p. 202). The investigators found that aspects of moral agency (e.g., awareness and acceptance of responsibilities including advocacy), moral imagination (e.g., empathy, clarifying patient preferences), and moral community (e.g., supporting moral communication and good relationships) protected these nurses from moral distress and its harmful consequences. The authors highlight the importance of learning more about how clinicians understand their moral identities, relationships, responsibilities, and values.

Studies that examine upstream factors such as moral obligations, identities, agency, accountabilities, communities, and networks could help illuminate strategies to promote aspects of clinicians' moral work that might prevent moral distress from occurring in everyday clinical practice. Researchers also need to develop and validate sound instruments for measuring these important moral concepts. Additionally, applying the strength-based, qualitative technique of appreciative inquiry that explores research phenomenon from a positive versus problem-based approach could be productive. Nurses, physicians, patients, families, other healthcare providers, and healthcare leaders could offer new insights on decreasing threats to moral integrity and mitigating moral distress through appreciative inquiry approaches regarding the meaning of moral integrity, moral practice, and moral communities. The steps of appreciative inquiry include (a) seeing the best of what is, (b) imagining the ideal, (c) designing what could be, and finally (d) creating the desired changes [34]. Appreciative inquiry research could help in developing policies and practices that support moral integrity and improve moral communities. These practices could then be evaluated.

7.3.3 Expanding the Study of Moral Distress

Researchers have primarily explored moral distress as a concept and experience of nurses, physicians, and other members of the healthcare team. Not much is known about moral decision-making processes among patients, families, and surrogates as they confront difficult treatment decisions in the context of uncertain prognosis, unknown outcomes, and often inadequate or inaccurate knowledge about the patient's condition while also experiencing strong emotions associated with caring for an ill family member (see essay by Mooney-Doyle, Chap. 3).

Cultural factors that influence patient, family, and surrogate decision-making also need more study.

Research also needs to expand understanding of team-based experiences and interventions in ethically difficult situations. In 2010, Ulrich, Hamric, and Grady suggested that ethics dialogue among healthcare team members could decrease moral distress [35]. The authors noted, "Moving to more open collegiality and shared practice models may help to alleviate moral distress by increasing a sense of shared responsibility and of professional satisfaction" (p. 21). The need for collaboration is illustrated in a pilot study that investigated the feasibility of using a screening tool to assess risk factors for ethical conflicts. The top three risk factors that emerged during the 3-month study period included patient suffering, family's unrealistic expectations, and nurses' moral distress [36]. We discussed this triad as "shared suffering" and noted that "when suffering is viewed solely as an individual experience, which it often is, feelings of isolation and being trapped in silence can result" (p. 254). However, if suffering is viewed as a shared experience in ethically difficult situations, then easing suffering could also include relational approaches designed to promote trust and effective team communication and decision-making.

However, very little is known about shared suffering and the facilitators and barriers of shared practices (such as shared decision-making) on individual and team-based moral integrity and distress. A qualitative study by Bruce and colleagues explored moral distress in critical care teams and found that intra-team discordance, particularly about situations involving non-beneficial treatments or full disclosure, was a key source of moral distress [37]. Investigators unveiled specific maladaptive behaviors and concluded that communication gaps especially at critical time-points compromise decision-making. Uninterrupted, these outcomes compound ethical conflicts which are known to diminish patient safety and care quality and increase the risk for moral distress. Learning more about the factors that promote effective team dynamics and assist teams to work effectively through moral disagreements seems essential. Important areas for research include studying the impact of interprofessional ethics education, regularly scheduled ethics team discussions, and shared clinical practice models on the experiences of patients, families, and clinicians, on specific patient outcomes and on system efficiency.

7.3.4 Examining the Concept of Moral Community

Creating moral spaces for interprofessional dialogue and reflection is critical to enhancing a moral community [24, 33, 38]. The health professions arose out of a need in society for care of human persons; in nursing, the social contract to those whom we serve is outlined in the American Nurses Association (ANA) social policy statement [39]. The needs of patients, both ill and well, are best addressed by interprofessional teams. Thus, any kind of moral community requires an interprofessional team, and functional teams are clearly a primary prevention strategy against the development of moral distress.

Pavlish and colleagues conducted an ethnographic study that explored ethical conflicts and the meaning of moral community in oncology [40]. Thirty nurses and twelve key informants such as clinical ethicists and physicians described ethically difficult situations, factors that contributed to ethical conflicts, and strategies that addressed these conflicts. Postponing or avoiding difficult conversations, suppressing different moral perspectives, and experiencing the tension between competing obligations were common. These experiences often escalated normal and anticipated moral disagreements into ethical conflict which then compromised relationships and patient care. Leadership that recognized the importance of building a moral community was key to de-escalating conflict. Strategies such as promptly addressing ethical issues through healthy dialogue and utilizing competent ethics resources created the necessary time and space for developing respectful interprofessional relationships and a sense of moral community. Investigations that examine ways to create a receptive ethical climate, cultivate a relational response to moral differences, and foster ethics accountability offer hope for improving care especially at the end of life when ethical issues tend to escalate.

7.3.4.1 Impact of Moral Spaces: Ethics Rounds to Promote Moral Community

One example of creating "moral space" to lessen moral distress is unit-based "ethics rounds" which have been implemented at several hospitals across the country. In our experience, "ethics rounds" can positively impact moral distress, yet to our knowledge, systematic study of this impact has not yet been undertaken. Most often, unit clinical nurses, in conjunction with other health professionals, ethics experts, and physicians, plan these rounds. Often, an "ethics expert" is present in the rounds, either as facilitator or as participant providing expert commentary. It is critical that nurses and others who are part of the healthcare team be involved in leading and planning the rounds because of their firsthand knowledge about patient situations and their ongoing collaborative efforts to improve each situation. The unit-based rounds can be case-based, topic-based, or theme-based. Nurse directors support the regular occurrence and attendance at these monthly, bimonthly, or quarterly rounds. Clinical nurses with demonstrated interest and continuing education in healthcare ethics organize and lead the rounds and are supported in preparation and in real time by nurse ethicists. Tools and tips for facilitating ethics rounds represent one concrete strategy for how clinical nurses and others are supported as facilitators. On one 19-bed medical-surgical unit which has been holding ethics rounds for over 5 years, 26 nurses (72% response rate) responded to a six-item survey about the utility of ethics rounds and its impact on their moral distress (Box 7.1). Research to learn which internal and external factors influence level of moral distress is needed along with the impact of ethics rounds in patient and/or surrogate outcomes such as satisfaction with communication and participation in treatment decision-making. Initiatives such as "compassionate care rounds" or "resiliency rounds" intended for clinicians to debate, debrief, and listen to one another and the extent to which these mitigate moral distress, affect patient experiences, influence teamwork, and contribute to a moral community should be studied.

Box 7.1 Quality Improvement Project on Ethics Rounds: Survey Results (N = 26)

Questions	Percent agreement (%)
1. Do "ethics rounds" on your unit enhance your moral agency in your professional role?	82
2. Do "ethics rounds" on your unit lessen the degree of moral distress for you?	70.9
3. As a result of ethics rounds, have you found that you approach a case/topic/practice in a new or different way?	81
4. As a result of ethics rounds, have you found that you approach a case/topic with a greater degree of confidence?	86.8
5. Do ethics rounds provide an opportunity to share your views as well as consider the views of others?	92.3
6. Do ethics rounds provide a safe space for you to participate in a discussion regarding a case or topic with ethical complexity?	86.3

Addendum
One nurse participant in the quality improvement survey stated that she is a better advocate for patients and families, knowing that she has the support of nursing leadership on the unit. Another nurse stated that ethics rounds have empowered him to have difficult conversations, and to question a patient's plan of care with a physician. We conclude from this quality improvement survey that ethics rounds may be one tangible strategy to support moral community in health care institutions

7.4   A Research Agenda for Action on Moral Distress

Empirical research is lacking around interventions intended to prevent or mitigate the effects of moral distress. Figure 7.1 illustrates at least three potential action points: prevention, risk reduction, and mitigation or treatment strategies. Sample studies and research recommendations follow.

7.4.1   Primary Prevention Strategies for Moral Distress

Research to develop and evaluate the impact of interventions that potentially prevent moral distress is needed. Current initiatives that should be studied and are described below include a moral courage policy, collaborative governance structures, and policy development approaches for patients at the end of life and for the chronically critically ill.

7.4.1.1   Speaking Up Policy
One initiative that could be studied as a possible prevention for moral distress has occurred in a freestanding children's hospital. Nurses developed a policy which sets a standard that staff members are expected to speak up in response to ethical or safety concerns in spite of perceived risks. At the same time, the standard requires

all staff to listen respectfully and respond collaboratively to safety and ethics concerns raised by others. This initiative also offers constructive approaches and resources that staff can apply as they act on their professional and moral responsibilities to help maintain moral integrity. That hospital's expectation is that staff will speak freely about ethical concerns and, in turn, can trust existing structures to support their exercise of moral courage, thereby decreasing moral distress. A policy so directly aimed at mitigating moral distress deserves study regarding its impact and effectiveness [41].

7.4.1.2 Collaborative Governance: Ethics in Clinical Practice Committee

At a northeastern quaternary care center, the chief nurse and senior vice president for patient care designed and implemented a professional practice structure that includes a "collaborative governance" communication and decision-making structure in 1997. Unlike many hospitals that with good intentions implement similar structures, yet find it challenging to maintain such, this institution's nursing leadership has maintained and built upon its structure, which is now 20 years strong [42]. With ongoing mentoring, clinicians in direct patient care roles lead several committees within this structure, including the "Ethics in Clinical Practice Committee" (EICPC) which brings together direct care providers from across the organization. During these meetings, clinicians share ethically challenging experiences in their practice; learn the language of ethical discourse; learn how to teach clinicians, patients, and families about advance care planning; and make policy recommendations which can positively impact ethical care in the organization. A grounding principle of this structure is that it is co-chaired by direct care clinicians, usually one clinical nurse and one clinician from an allied health discipline. These co-chairs are supported by a leadership coach, who has ethics expertise and knowledge of how to guide an issue through the organization. Box 7.2 offers a recent example of how this governance committee utilizes its structure to develop policy and consider implications. Evaluation of the impact of collaborative governance structures while challenging has been attempted [43], although it is not yet known how such collaborative or shared governance committees impact clinicians' experiences of moral distress. Research is also needed to determine how such collaborative structures impact clinicians' moral agency and engagement, job satisfaction and retention, as well as patient outcomes. Much research remains.

Box 7.2 Preventing Moral Distress: Collaborative Governance
Members were asked to bring case exemplars or to reflect on "what keeps them awake at night." These situations generally highlighted areas for improvement in ethical care. A nurse brought forth a case related to "DNR suspension in the OR"—she described that some physicians state that DNR

must always be reversed when a patient goes to surgery or a procedure. The Co-Chair of the Ethics in Clinical Practice committee, a clinical nurse on a medical floor, investigated the clinical nurse identified problem. A policy section in the organization's Life Sustaining treatment Policy that was not well known to practicing nurses was uncovered. This policy section revealed that which is consistent with current professional position statements from American College of Surgeons, AORN, Nurse Anesthetists' and Anesthesiologists' Societies, which is to support an individualized approach to policies and practice around DNR in OR/Procedures.

A panel was formed to attend the Ethics in Clinical Practice committee meeting. This panel included the Director of Surgical ICU (anesthesiologist), GI physician from interventional radiology who had raised ethical concerns about the practice of reversing DNRs for certain groups of patients, leadership of Optimum Care (Ethics) Committee which oversees Life Sustaining Treatment Policies, an ICU nurse who was an EICPC member and could speak to practice in this area. A robust discussion that drew on professional positions statements, professional literature, hospital policy, and case applications occurred.

The outcomes included: (a) EICPC members returning to their units and surveying interprofessional clinicians, including nurses, respiratory therapists, attending and resident physicians to assess their knowledge of the policy along with its professional association support, and (b) teach accordingly to support a patient centered model in decision making about "DNR suspension or not" in the OR/Procedure.

7.4.1.3 Policy Development for Patients Approaching End of Life

Clinical nurses, particularly in acute care settings, practice in both independent and collaborative roles [44, 45]. In their collaborative role, which in ICUs can be up to 85–90% of practice, clinical nurses continuously assess patients with multisystem disease who are receiving complex medical interventions including but not limited to mechanical ventilators, vasoactive drugs, extracorporeal membrane oxygenation (ECMO), and intracranial pressure monitoring (ICP). Nurses attend patients and their families around the clock and monitor the patient's response to disease and its interventions. While rendering prognosis is not officially recognized as within the professional nurse's role, experienced clinical nurses have a great deal of insight, drawn from objective measures as well as subjective reports from patients and their families, coupled with clinical nursing wisdom, regarding the benefits and burdens of ongoing disease and treatment. Research has documented that unaddressed end-of-life issues are significant contributors to nurses' moral distress [13, 14, 46]. Less well studied is the identification of burdensome treatment as ethically problematic for patients, families, and clinicians at large and how nurses' voices contribute and how a unit or hospital's ethical climate can contribute to nurses' voices being heard.

For example, in individual cases, as well as in the identification of themes that lead to organizational policy change, it has been our experience that clinical nurses who call the team's attention to ethically problematic cases such as "full code status" for a patient who is imminently dying can impact patient outcomes. Facilitating frank discussion on goals of care can lead to a peaceful death for the patient and possibly impact the bereavement of family in a positive way. It is well documented that the role of surrogate decision-maker is a burdensome role [47], and families may experience relief when physicians and other members of the team compassionately guide this decision. Studying the impact of policies that include nurses' roles in facilitating shared decision-making seems essential for quality end-of-life care.

At one quaternary care institution, clinical nurses were very much involved in discussions and in crafting policy language that would protect patients on a trajectory toward death from the harms of cardiopulmonary resuscitation when their surrogates were unable to agree to a "do not resuscitate" order. Researchers at this institution which includes clinical nurses as part of the research team provide empirical support for a "medically indicated DNR" as end-of-life approaches. In this institution, research has demonstrated that surrogates, for the most part, accept the change to DNR and seem to be able, with nursing support, to redirect themselves to the bedside of their loved one [48].

Additionally, nurse involvement in policy writing that protects patients from harm can impact compassionate implementation of such policies, in which the patient can be protected from cardiopulmonary resuscitation at the end of life, while the surrogate decision-maker, no matter what their psychosocial-spiritual needs or motivation, is also held in compassion during end-of-life care [49]. One would hypothesize that having such institutional policies as described in Box 7.3 can lessen the moral distress of nurses, house officers, and respiratory therapists, whose hands are first on deck to provide resuscitation. However, to our knowledge, measurement of such impact has not yet been studied.

Box 7.3 Sample Institutional Policy

Doing No Harm—the responsible physician always has an overriding responsibility to protect the patient from harm. In some clinical situations, the responsible physician may determine, after exploring and documenting a patient's values and beliefs, and in conjunction with clinicians involved in the patient's care, that attempting cardiopulmonary resuscitation would be more harmful than beneficial for the patient. In such situations, the responsible physician may decide not to offer CPR. In two such situations, the responsible physician may follow guidelines for entering appropriate code status orders, which may include do not resuscitate ("DNR"), do not intubate ("DNI"), or both.

- *Situation 1*: The responsible physician should consider protecting a patient who is imminently dying from CPR's potential harms by not offering CPR and entering the appropriate orders. In this situation, the responsible physician may decide, but is not required, to obtain a second opinion about not offering CPR from another senior or experienced physician or from the OCC and also may request advice from the Office of General Counsel ("OGC").

- *Situation 2*: The responsible physician may also consider not offering CPR to a patient who is not imminently dying but who has no reasonable chance of surviving CPR to the point of leaving the hospital. In this case, if, after careful discussion with the patient or surrogate, the patient or surrogate does not assent to the plan, orders to withhold CPR should be entered only if another senior or experienced physician and a consultant from the OCC concur with the plan and only if this concurrence has been documented in the medical record.
- In either circumstance, the responsible physician who decides not to offer CPR should inform the patient or surrogate of this decision and its rationale and assure that the patient will continue to receive the highest possible quality of care.

Excerpt from Massachusetts General Hospital Life Sustaining Treatment Policy

7.4.1.4 Proactive Collaboration for Patients Who Are Chronically Critically Ill

Increasingly, life-sustaining treatments like tertiary antibiotics, mechanical cardiac support, and continuous venous-venous hemofiltration (CVVH) in intensive care units, stepdown units, and chronic care facilities can prolong life. Health systems consider such patients as the "chronically critically ill" [50]. Nurses, as around-the-clock professional caregivers, witness suffering as patients, who are often without a voice to determine their own fate, respond to treatments [51]. Given their collaborative role with physicians, as implementers and monitors of medical interventions, yet, without autonomy over medical decision-making, nurses at times have been caught in the quandary of continuing painful treatments with limited or no benefit for patients. Moral distress may result.

As the population of chronically critically ill patients grows, clinicians should collaborate to craft policies that support patient and surrogate education and advance care planning across clinic, long-term, and acute care settings with regularly scheduled conversations on options and implications of treatment decisions. Patients and families appear to benefit from advance directives [52]. However, in a systematic review, authors found among 795,909 patients, only 36.7% had completed an advance directive [53]. Although this study did not target just chronically ill patients, at least one study showed that most patients with chronic critical illness have not appointed a surrogate or expressed preferences regarding life-sustaining treatments

[54]. While such patients may not be imminently dying, prognosis often suggests that many of these patients will never be able to live independently. Educative and supportive decision-making strategies to assist patients, families, and surrogates in identifying treatment preferences along the trajectory of chronic critical illness should be developed and evaluated. Longitudinal research can assess the effectiveness of these collaborative, across-setting strategies on patient and family decision-making and satisfaction with care and the intensity of clinician moral distress. The impact on the healthcare system should also be evaluated.

7.4.2   Risk Reduction Strategies for Moral Distress

Research on educational programs and system-level interventions that demonstrate a positive impact in reducing the risk of developing moral distress is essential. Some examples follow.

7.4.2.1 Ethics Early Action Protocol

Chapter authors recently investigated moral distress levels, among other clinician outcomes, before and after implementation of an intervention designed to stimulate early action in response to ethical issues. The intervention included three parts: an Ethics Early Action Protocol (EEAP), an online-training program for providers to apply the protocol, and an ethics app for ongoing ethics support.

EEAP focuses the healthcare team's attention on dimensions of the patient situation and care plan that are associated with clinical ethical concerns. First, the protocol consolidates common ethics demographic information like named decision-maker, presence of advance care planning documents, and code conversations for the individual patient. Then, based on studies that explored nurse and physician perspectives about common risk factors that contribute to ethics issues, EEAP requires clinicians to identify patient, family, and situational risk factors (including critical risk factors in each category) that pertain to the situation being assessed (Box 7.4) [55, 56]. For example, the team assesses the patient circumstances for risk factors and then adds up the factors at play to judge if there is a low, medium, or high risk for ethical conflict. Finally, suggested interventions are offered based on the assessed risk, for example, "talk with nursing leadership" or "call an ethics consult."

Box 7.4 Sample Risk Factors for Ethical Conflicts
Sample of patient risk factors:

- Potential for escalation of nonbeneficial treatment
- Combination of patient lacking decisional capacity and family conflict
- Patient suffering (i.e., high anxiety or physical, psychological, or spiritual pain)

Sample of family risk factors:

- Conflict between family members about plan of care
- Cultural or faith beliefs that influence family expectations
- Family absent or unavailable

Sample of situational risk factors

- Need for conversation about goals of care and/or cohesive plan of treatment
- Compromised trust
- Signs of moral distress in families and/or clinicians

EEAP was designed for use by the interdisciplinary team in intensive care units as a standardized, proactive prompt for team-based conversations about the ethical dimensions of care. It was implemented on six different intensive care units (two in each of three major academic medical centers) with expectations that an ethics assessment would be conducted on every patient every day. Prior to implementation of the protocol, clinician study participants (nurses, physicians, and advance practice practitioners) were surveyed for evidence of (1) moral distress using the Moral Distress Scale-Revised (MDS-R); (2) ethics self-efficacy, using an ethics self-efficacy scale (ESES); and (3) perceptions of ethical climate with the Hospital Ethical Climate Survey (HECS). The surveys were repeated at 3 and 6 month post-implementation.

All three clinician-related outcomes–moral distress, ethics self-efficacy, and hospital climate–showed statistically significant changes over the course of the study period. Moral distress decreased while clinicians' perception of the ethical work environment and perception of their own effectiveness in acting in ethically complex situations increased. The study builds evidence for the usefulness of ethics-specific assessment and proactive ethics interventions to enhance resilience and prevent or mitigate the effects of moral distress. It supports the idea that ethics deliberation and interdisciplinary communication about immediate or impending ethical concerns could improve quality of care and sanction ethics-based dialogue as an essential aspect of patient-centered care. Other such communication protocols, early identification techniques, and team-based strategies need development and investigation for their impact on patient, family, clinician, and healthcare system outcomes.

7.4.2.2 Clinical Ethics Residency for Nurses

A clinical ethics residency for nurses at Massachusetts General Hospital, supported by a grant from Health and Human Resources Division of Nursing Bureau of Health Professions, was a collaborative among nurse ethicist faculty and a clinical pastoral education supervisor from two acute care hospitals and a local college. Practicing clinical nurses and nurses in leadership programs were invited to apply, making participation approximately 70% clinical nurses and 30% nursing education and leadership. Application essays included situations replete with moral distress [57]. A 3-year program allowed three cohorts of clinical nurses, nurse leaders, and faculty to participate in the residency [58]. Attendance of clinical nurses at this 10-month long, 98-hour program was supported by nurse directors, who were in turn supported by the chief nurse. Pre- and post-evaluation measures with MDS-R and an ethics self-efficacy scale showed statistically significant results for decreased moral distress [57]. Nurses who attended this program remain active today in their respective units in such roles as facilitators of unit-based ethics rounds and as hospital ethics committee members, all of which have impacted ethical practices in their units or in other venues throughout the hospital. This initiative could be replicated in additional settings and should undergo comparative effective testing with other initiatives.

7.4.3   Mitigation and Treatment Strategies for Moral Distress

Developmentally as one might expect, research began by defining and measuring moral distress and subsequently exploring its effects on patient care quality and safety, clinician well-being, and organizational cost. More recently, researchers have begun to investigate mitigation efforts. Given that the very nature of ethical dilemmas will almost certainly at times result in moral distress and its residue, research should aim to explore treatment strategies for clinicians experiencing moral distress.

7.4.3.1 Moral Distress Consultation Service

Hamric and Epstein report on the development and evaluation of a Moral Distress Consultation Service (MDCS) [14]. As part of an ethics consultation service, MDCS responds to clinicians' requests to discuss a clinical situation that is causing moral distress. Two moral distress consultants elicit a description of the situation and assist the healthcare team to negotiate differences, identify what most participants believe is the "right" action, and pinpoint barriers to taking action. The consultants focus on professional standards, values, and obligations to guide participants' analysis of the situation. Finally, once participants reach consensus on the "right" action, consultants help clinicians strategize to remove barriers to the selected action. Early and ongoing evaluation indicates that MDCS not only empowers clinicians; it also encourages engagement, collaboration, and system-wide changes. Studies that measure patient, clinician, team, and organizational outcomes are needed.

7.4.3.2 Healing from Moral Distress

Research on mentorship programs that develop healthcare provider moral resilience, ethics preparedness, and ethical competence is important. Berggren and Severinsson's research examined the effect of clinical supervision and mentorship for nurses who work in complex clinical situations [59]. Small group sessions over two semesters appeared to impact nurses' moral decision-making. Post-session interviews indicated that mentorship increased nurses' self-confidence, sense of accountability, and ability to establish meaningful and supportive relationships with patients. Furthermore, authors noted that the healing aspect of these sessions resulted from the nurses' relationship with the clinical supervisor/mentor who provided an opportunity for nurses to share their narratives about difficult situations while offering guidance to support moral decision-making.

A Canadian initiative led by Dr. Peter Dodek is currently evaluating the effectiveness of an action-based moral conflict assessment (MCA) process created by Chavelier and Buckles [60]. Building on the contributions of several disciplines, this process assumes that moral stress, suffering, and distress may reflect or even hide other issues that need to be uncovered and considered. Values, interests, and identities combine in various ways during clinical situations with the implication that each moral distress situation must be "worked through" and assessed on its own by those involved. The MCA process not only assists teams to consider all aspects of a difficult situation, especially those that are not immediately apparent such as assumptions, it also reveals individual and team-based coping strategies. When applying the MCA intervention, clinicians begin to recognize individual and collective moral agency along with opportunities for learning and growth which are a part of every moral distress situation [61]. This team-based intervention, which requires further evaluation and replication, may provide an opportunity for healing from moral distress.

Another program that holds promise for restorative outcomes is a 4-day writing retreat for clinicians experiencing moral distress [62]. Writing and storytelling have been effective in healing both patient populations and healthcare professionals [63–65]. The writing retreat offers didactics on moral distress and illness narratives accompanied by dynamic, exploratory writing assignments, designed to transform moral distress experiences from rumination and worry into a more reflective space. For example, one assignment utilized a forgiveness writing exercise to address clinician regret and self-blame. The program also integrated movement therapy, which acts synergistically with writing to heal trauma [66].

The retreat provided a safe space for 20 interdisciplinary healthcare professionals and demonstrated important outcomes. Pre-retreat, nearly 40% of the group acknowledged dealing with troubling ethical issues daily to weekly, with 61% rating their moral distress as moderate to high. Post-retreat evaluations indicated that participants increased their awareness of moral distress among clinicians of all disciplines. Themes that emerged in the evaluations were connection to others, increased coping mechanisms, and decreased isolation. During the retreat, 65% of participants experienced vulnerability with about half of that group experiencing more and the other half experiencing less vulnerability than before the retreat; yet all described the experience positively in terms of growth, empowerment, trust, and/or strength. The 3-month post-evaluation (65% response rate) showed a sustained effect; nearly 70% were still writing and indicated that feeling empowered to speak up decreased their stress level. Movement therapy during the retreat made the group more aware of how they embodied moral distress, and this awareness was sustained 3-month post-retreat for 85% of the responders. Overall, there was a stronger sense of voice. One participant stated that the retreat "allowed me to make my thoughts and feelings public. I have been privileged to listen to the distress and sadness of others. To be a witness has opened my soul." This program would clearly benefit from replication as a research study to determine its effect on patient and clinician outcomes.

Future research should also investigate the use of art, theater, dance, exercise, and other activities as potential modalities and integrate the relational aspects that not only create individual healing but also possibly secondary effects of creating moral community. Existing renewal and self-care programs should also be evaluated for their impact on moral resilience. The work of Davidson, Agan, and Chakedis on blame-related moral distress demonstrates the importance of developing and testing methods to promote clinician healing from moral distress [67].

A few studies have explored family conferences as a means to address family and/or surrogate distress some of which may be moral distress [47, 68–74]. Unfortunately, very little is known about moral distress among patient, family, or surrogate decision-makers—nor is there enough preliminary work to develop an instrument that specifically measures moral distress among stakeholders other than direct care providers. Related research points to the importance of the topic. For example, Braus and colleagues examined the effect of proactive palliative care rounding on the likelihood and timing of family conferences and the presence of post-traumatic stress disorder (PTSD) among family respondents [75]. The intervention increased the number of family conferences (which also occurred sooner) and decreased family members' PTSD symptoms. However, in a different study [76] with family members of patients with chronic critical illness, proactive palliative care discussions did not reduce depression and anxiety symptoms and may have increased PTSD. These differences indicate the need for more in-depth investigations about moral decision-making and moral distress among patients, family, and surrogates. Studies that develop valid instruments and effective interventions should follow. Interventions that promote patient and family resilience as well as surrogate decision-making should also be developed and tested. Finally, given the emerging literature [77] on trust and team performance, interventions that increase trust between patients, families, and clinicians should be developed and tested for effectiveness in facilitating family decision-making and mitigating patient and family moral distress.

7.5 Final Thoughts on Researching Moral Distress

This call for research, issued 33 years after Andrew Jameton defined moral distress, reminds us of the impasses and obstacles that arise, perhaps even blind us from what is deeply needed in building moral distress science, that is, to be genuinely curious

Table 7.1 Assessment device for moral distress research

Population	Level I Systematic Review and RCT Meta-Analysis [78]	Level 2 RCT	Level 3 Controlled trial (no randomization)	Level 4 Case Control or Cohort Study	Level 5 Systematic review of descriptive, qualitative studies	Level 6 Single descriptive or qualitative study
Patient						
Family						
Populations						
Clinicians in specific settings						
Healthcare Team						
Organizational systems and processes						
Specialty practices						
Political and social contexts						

and open, work out differences collaboratively and systematically, invite all voices (especially those with new ideas or those previously unnoticed), and appreciate the depth of healing and the expanse of creative ingenuity that will be required. We as clinicians interested in advancing moral distress science need to absorb the solid foundation that exists and move to cohesively and systematically address the challenges issued by the Quadruple Aim [8] and the critical care organizations' Call to Action [6] that now recognize the critical need to address moral distress and clinician well-being. Tabulating accumulating evidence (see sample Table 7.1) will reveal important gaps and strengths for moral distress researchers. State-of-the-science articles [13] and systematic reviews [30] will facilitate our collaborative assessment.

A research agenda will collapse without financial support; moral distress is currently underfunded. Clinicians and ethicists who work in complex clinical situations embedded in multi-faceted health eco-political systems, must provide convincing evidence to funding agencies and philanthropists that resources for moral distress research will make a difference at all levels of the organization, in all spheres of the moral community, and to the public's health. The demands are many, yet as we concentrate on our collective obligations, invite creative ideas, and blend many voices to collaborate in research across disciplines, clinical environments, the academy, and the public, synergy will arise for the challenge before us.

> We must be still and still moving, into another intensity. For a further union, a deeper communion. (Eliot [79])

References

1. Epstein E. Moral obligations of nurses and physicians in neonatal end-of-life care. Nurs Ethics. 2010;17(5):577–89.
2. Cronenwett L, Sherwood G, Barnsteiner J, Disch J, Johnson J, Mitchell P, Sullivan DT, Warren J. Quality and safety education for nurses. Nurs Outlook. 2007;55(3):122–31.
3. Institute of Medicine (US) Committee on the Health Professions Education Summit; Greiner AC, Knebel E, editors. Health professions education: a bridge to quality. Washington DC: National Academies Press; 2003. Chapter 3, The Core Competencies Needed for Health Care Professionals. https://www.ncbi.nlm.nih.gov/books/NBK221519/.
4. Hyatt J. Recognizing moral disengagement and its impact on patient safety. J Nurs Regul. 2017;7(4):15–21.
5. American Association of Critical-Care Nurses. The 4 A's to Rise Above Moral Distress. http://www.aacn.org/moraldistress4As. Accessed July 17, 2017.
6. Moss M, Good VS, Gozal D, Kleinpell R, Sessler CN. An official critical care societies collaborative statement—burnout syndrome in critical care health-care professionals: a call for action. Chest. 2016;150(1):17–26.
7. Berwick DM, Nolan TW, Whittington J. The triple aim: care, health, and cost. Health Aff. 2008;27(3):759–69.
8. Bodenheimer T, Sinsky C. From triple to quadruple aim: care of the patient requires care of the provider. Ann Fam Med. 2014;12(6):573–6.
9. National Academy of Medicine. Action collaborative on clinician well-being and resilience. 2017. Retrieved from https://nam.edu/initiatives/clinician-resilience-and-well-being/. Accessed July 3, 2017.
10. Lützén K, Kvist BE. Moral distress: a comparative analysis of theoretical understandings and inter-related concepts. HEC Forum. 2012;24(1):13–25.
11. Varcoe C, Pauly B, Storch J, Newton L, Makaroff K. Nurses' perceptions of and responses to morally distressing situations. Nurs Ethics. 2012;19(4):488–500.
12. Epstein EG, Hamric AB. Moral distress, moral residue, and the crescendo effect. J Clin Ethics. 2009;20(4):330.
13. McAndrew NS, Leske J, Schroeter K. Moral distress in critical care nursing: the state of the science. Nurs Ethics. 2016. https://doi.org/10.1177/0969733016664975.
14. Hamric AB, Epstein EGA. Health system-wide moral distress consultation service: development and evaluation. HEC Forum. 2017;29(2):127–43.
15. Hamric AB, Wocial LD. Institutional ethics resources: creating moral spaces. Hastings Cent Rep. 2016;46:S1.
16. Grzywacz JG, Fuqua J. The social ecology of health: leverage points and linkages. Behav Med. 2000;26(3):101–15.
17. Trickett EJ. Toward a framework for defining and resolving ethical issues in the protection of communities involved in primary prevention projects. Ethics Behav. 1998;8(4):321–37.
18. Krieger N. Theories for social epidemiology in the 21st century: an ecosocial perspective. Int J Epidemiol. 2001;30(4):668–77.
19. Rathert C, May DR, Chung HS. Nurse moral distress: a survey identifying predictors and potential interventions. Int J Nurs Stud. 2016;53:39–49.
20. Peter E. Guest editorial: three recommendations for the future of moral distress scholarship. Nurs Ethics. 2015;22(1):3–4.
21. Campbell SM, Ulrich CM, Grady C. A broader understanding of moral distress. Am J Bioeth. 2016;16(12):2–9.
22. Fourie C. Moral distress and moral conflict in clinical ethics. Bioethics. 2015;29(2):91–7.
23. Jameton A. Nursing practice: the ethical issues. Prentice Hall: Engelwood Cliffs; 1984.
24. Peter E, Liaschenko J. Moral distress reexamined: a feminist interpretation of nurses' identities, relationships, and responsibilities. J Bioethical Inq. 2013;10(3):337–45.
25. Hamric AB, Borchers CT, Epstein EG. Development and testing of an instrument to measure moral distress in healthcare professionals. AJOB Primary Res. 2012;3(2):1–9.

26. Wocial LD, Weaver MT. Development and psychometric testing of a new tool for detecting moral distress: the moral distress thermometer. J Adv Nurs. 2013;69(1):167–74.

27. Hamric AB. Empirical research on moral distress: issues, challenges, and opportunities. HEC Forum. 2012;24(1):39–49.

28. Henrich NJ, Dodek PM, Alden L, Keenan SP, Reynolds S, Rodney P. Causes of moral distress in the intensive care unit: a qualitative study. J Crit Care. 2016;35:57–62.

29. Huffman DM, Rittenmeyer L. How professional nurses working in hospital environments experience moral distress: a systematic review. Critical Care Nurs Clin. 2012;24(1):91–100.

30. Lamiani G, Borghi L, Argentero P. When healthcare professionals cannot do the right thing: a systematic review of moral distress and its correlates. J Health Psychol. 2017;22(1):51–67.

31. Oh Y, Gastmans C. Moral distress experienced by nurses: a quantitative literature review. Nurs Ethics. 2015;22(1):15–31.

32. Pauly BM, Varcoe C, Storch J. Framing the issues: moral distress in health care. HEC Forum. 2012;24(1):1–11.

33. Traudt T, Liaschenko J, Peden-McAlpine C. Moral agency, moral imagination, and moral community: antidotes to moral distress. J Clin Ethics. 2016;27(3):201–13.

34. Cooperrider DL, Whitney DK, Stavros JM. Appreciative inquiry handbook. San-Francisco: Berrett-Koehler Publishers; 2003.

35. Ulrich CM, Hamric AB, Grady C. Moral distress: a growing problem in the health professions? Hastings Cent Rep. 2010;40(1):20–2.

36. Pavlish CL, Hellyer JH, Brown-Saltzman K, Miers AG, Squire K. Screening situations for risk of ethical conflicts: a pilot study. Am J Crit Care. 2015;24(3):248–56.

37. Bruce CR, Miller SM, Zimmerman JLA. qualitative study exploring moral distress in the ICU team: the importance of unit functionality and intrateam dynamics. Crit Care Med. 2015;43(4):823–31.

38. Walker MU. Moral understandings: a feminist study in ethics. New York: Oxford University Press; 2007.

39. American Nurses Association. Nursing's social policy statement: The essence of the profession. 2010. https://www.Nursesbooks.org.

40. Pavlish C, Brown-Saltzman K, Jakel P, Fine A. The nature of ethical conflicts and the meaning of moral community in oncology practice. Oncol Nurs Forum. 2014;41(2):130–40.

41. Fitzgerald H. Moral courage policy development in a pediatric setting: ethics liaisons influencing organizational ethical climate. Houston: ASBH meeting (Ethics and Creative Expression); 2015.

42. Erickson JI, Jones DA, Ditomassi M. Fostering nurse-led care: professional practice for the bedside leader from Massachusetts General Hospital. Indianapolis: Sigma Theta Tau International; 2013.

43. Erickson JI, Hamilton GA, Jones DE, Ditomassi M. The value of collaborative governance/staff empowerment. J Nurs Adm. 2013;33(2):96–104.

44. Gordon M. Nursing diagnosis: process and application. New York: McGraw Hill; 1982.

45. Grace PA, Robinson EM. Nursing's moral imperative (chapter 6). In: Fostering nurse-led care: professional practice for the bedside leader from Massachusetts General Hospital. Indianapolis: Sigma Theta Tau International; 2013. p. 123–52.

46. Mealer M, Moss M. Moral distress in ICU nurses. Intensive Care Med. 2016;42(10):1615–7.

47. Davidson JE, Aslakson RA, Long AC, Puntillo KA, Kross EK, Hart J, Cox CE, Wunsch H, Wickline MA, Nunnally ME, Netzer G. Guidelines for family-centered care in the neonatal, pediatric, and adult ICU. Crit Care Med. 2017;45(1):103–28.

48. Courtwright AM, Brackett S, Cadge W, Krakauer EL, Robinson EM. Experience with a hospital policy on not offering cardiopulmonary resuscitation when believed more harmful than beneficial. J Crit Care. 2015;30(1):173–7.

49. Robinson EM, Cadge W, Zollfrank AA, Cremens MC, Courtwright AM. After the DNR: surrogates who persist in requesting cardiopulmonary resuscitation. Hastings Cent Rep. 2017;47(1):10–9.

50. Lamas D. Chronic critical illness. N Engl J Med. 2014;370(2):175–7.

51. Gadow S. Existential advocacy: philosophical foundations of nursing. NLN Publ. 1990;20(2294):41–51.
52. Silveira MJ, Kim SY, Langa KM. Advance directives and outcomes of surrogate decision making before death. N Engl J Med. 2010;362(13):1211–8.
53. Yadav KN, Gabler NB, Cooney E, Kent S, Kim J, Herbst N, Mante A, Halpern SD, Courtright KR. Approximately one in three US adults completes any type of advance directive for end-of-life care. Health Aff. 2017;36(7):1244–51.
54. Camhi SL, Mercado AF, Morrison RS, Du Q, Platt DM, August GI, Nelson JE. Deciding in the dark: advance directives and continuation of treatment in chronic critical illness. Crit Care Med. 2009;37(3):919.
55. Pavlish C, Brown-Saltzman K, Hersh M, Shirk M, Nudelman O. Early indicators and risk factors for ethical issues in clinical practice. J Nurs Scholarsh. 2011;43(1):13–21.
56. Pavlish C, Brown-Saltzman K, Dirksen KM, Fine A. Physicians' perspectives on ethically challenging situations: early identification and action. AJOB Empirical Bioethics. 2015;6(3):28–40.
57. Robinson EM, Lee SM, Zollfrank A, Jurchak M, Frost D, Grace P. Enhancing moral agency: clinical ethics residency for nurses. Hastings Cent Rep. 2014;44(5):12–20.
58. Grace PJ, Robinson EM, Jurchak M, Zollfrank AA, Lee SM. Clinical ethics residency for nurses: an education model to decrease moral distress and strengthen nurse retention in acute care. J Nurs Adm. 2014;44(12):640–6.
59. Berggren I, Severinsson E. The influence of clinical supervision on nurses' moral decision making. Nurs Ethics. 2000;7:124–33.
60. Chevalier JM, Buckles DJ. Participatory action research: theory and methods for engaged enquiry. Abingdon: Routledge; 2013.
61. McCarthy J, Deady R. Moral distress reconsidered. Nurs Ethics. 2008;15:254.
62. Pennebaker JW. Telling stories: the health benefits of narrative. Lit Med. 2000;19(1):3–18.
63. Di Blasio P, Camisasca E, Caravita SC, Ionio C, Milani L, Valtolina GG. The effects of expressive writing on postpartum depression and posttraumatic stress symptoms. Psychol Rep. 2015;117(3):856–82.
64. Frank AW. The wounded storyteller: body, illness, and ethics. Chicago: University of Chicago Press; 2013.
65. Pennebaker JW, Smyth JM. Opening up by writing it down: How expressive writing improves health and eases emotional pain. New York: Guilford Publications; 2016.
66. Cancienne MB, Snowber CN. Writing rhythm: movement as method. Qual Inq. 2003;9(2):237–53.
67. Davidson JE, Agan DL, Chakedis S. Exploring distress caused by blame for a negative patient outcome. J Nurs Adm. 2016;46(1):18–24.
68. Curtis JR, Vincent JL. Ethics and end-of-life care for adults in the intensive care unit. Lancet. 2010;2010(376):1347–53.
69. Ellis L, Gergen J, Wohlgemuth L, Nolan MT, Aslakson R. Empowering the "cheerers": role of surgical intensive care unit nurses in enhancing family resilience. Am J Crit Care. 2016;25(1):39–45.
70. Iverson E, Celious A, Kennedy CR, Shehane E, Eastman A, Warren V, Freeman BD. Factors affecting stress experienced by surrogate decision makers for critically ill patients: implications for nursing practice. Intensive Crit Care Nurs. 2014;30(2):77–85.
71. Milic MM, Puntillo K, Turner K, Joseph D, Peters N, Ryan R, Schuster C, Winfree H, Cimino J, Anderson WG. Communicating with patients' families and physicians about prognosis and goals of care. Am J Crit Care. 2015;24(4):e56–64.
72. Netzer G, Sullivan DR. Recognizing, naming, and measuring a family intensive care unit syndrome. Ann Am Thorac Soc. 2014;11(3):435–41.
73. White DB. Rethinking interventions to improve surrogate decision making in intensive care units. Am J Crit Care. 2011;20(3):252–7.
74. You JJ, Downar J, Fowler RA, Lamontagne F, Ma IW, Jayaraman D, Kryworuchko J, Strachan PH, Ilan R, Nijjar AP, Neary J. Barriers to goals of care discussions with seriously ill hos-

pitalized patients and their families: a multicenter survey of clinicians. JAMA Intern Med. 2015;175(4):549–56.

75. Braus N, Campbell TC, Kwekkeboom KL, Ferguson S, Harvey C, Krupp AE, Lohmeier T, Repplinger MD, Westergaard RP, Jacobs EA, Roberts KF. Prospective study of a proactive palliative care rounding intervention in a medical ICU. Intensive Care Med. 2016;42(1):54–62.

76. Carson SS, Cox CE, Wallenstein S, Hanson LC, Danis M, Tulsky JA, Chai E, Nelson JE. Effect of palliative care–led meetings for families of patients with chronic critical illness: a randomized clinical trial. JAMA. 2016;316(1):51–62.

77. De Jong BA, Dirks KT, Gillespie N. Trust and team performance: a meta-analysis of main effects, moderators, and covariates. J Appl Psychol. 2016;101(8):1134.

78. Melnyk BM and Fineout-Overholt E. Evidence-based practice in nursing & healthcare: A guide to best practice. Wolters Kluwer Health: Philadelphia; 2015.

79. Bachelard G, Baird J, Bancroft A, Barthes R, Ackroyd P, Eliot TS. A life. New York: Simon and Schuster; 1984. p. 20.

International Perspectives on Moral Distress

Connie M. Ulrich, An Lievrouw, Bo Van den Bulcke,
Dominique Benoit, Ruth Piers, Georgina Morley,
Renatha Joseph, Baraka Morris, Subadhra D. Rai,
and Margaret Mei Ling Soon

C.M. Ulrich (✉)
Lillian S. Brunner Endowed Chair, University of Pennsylvania School of Nursing,
Philadelphia, PA, USA

Department of Medical Ethics and Health Policy, Perelman School of Medicine,
University of Pennsylvania School of Medicine, Philadelphia, PA, USA
e-mail: culrich@nursing.upenn.edu

A. Lievrouw
Cancer Centre, Ghent University Hospital, Ghent, Belgium

B. Van den Bulcke • D. Benoit
Department of Intensive Care Medicine, Ghent University Hospital, Ghent, Belgium

R. Piers
Department of Geriatric Medicine, Ghent University Hospital, Ghent, Belgium

G. Morley
Bioethics, University of Bristol, Centre for Ethics in Medicine, Bristol, UK

Critical Care Nurse, Barts Heart Centre, Barts Health NHS Trust, London, UK

R. Joseph
Department of Bioethics and Health Professionalism, Muhimbili University of Health
and Allied Sciences, Dar es Salaam, Tanzania

B. Morris
Department of Bioethics and Health Professionalism, Muhimbili University of Health
and Allied Sciences, Dar es Salaam, Tanzania

Nursing Department of Management, Muhimbili University of Health and Allied Sciences,
Dar es Salaam, Tanzania

S.D. Rai
School of Health Sciences (Nursing), Nanyang Polytechnic, Singapore, Singapore

M.M.L. Soon
Nursing Service, Tan Tock Seng Hospital, Singapore, Singapore

© Springer International Publishing AG 2018   127
C.M. Ulrich, C. Grady (eds.), *Moral Distress in the Health Professions*,
https://doi.org/10.1007/978-3-319-64626-8_8

8.1   International Perspectives on Moral Distress

Moral distress is not isolated to one geographical location. Indeed, health care clinicians on every medical, surgical, and intensive care hospital unit within the United States and abroad experience moral distress. Although much of the literature on moral distress has been discussed within the confines of the ethical issues and problems that American nurses, physicians and other types of clinicians encounter, moral distress may be at least as or even more serious for those who work in certain communities around the globe. And, moral distress is heightened for clinicians in some countries because the resources available to them are often scarce, testing clinicians' professional and moral obligations to their patients on a daily basis.

This chapter shares international perspectives on moral distress from physicians and nurses practicing in Africa, Asia, and Europe. In these essays, the accounts of moral distress are similar (in many ways) to the experiences of American clinicians. Themes across these essays include feelings of powerlessness, lack of voice, and limited autonomy or limited involvement in patient-related decisions. Perceptions of inappropriate care at end-of-life, particularly in European intensive care units mirrors Western concerns and begs for dialogue on ways to mitigate such distress within these specialty units of care. A British nurse expresses her lack of educational preparation to manage ethical issues in clinical practice and her sense of unease and uncertainty on how to morally navigate through these challenges, contributing to her moral distress. Our African colleagues tell a story of limited patient resources with the lack of blood in the local blood bank that would potentially save lives and the need for basic resources that support clinicians in their day-to-day work. Our Asian colleague tells a different type of story and shares her personal experience with truth-telling to a family member who is dying from cancer. In this essay, she poignantly stresses the moral distress that she felt as a daughter who was also a nurse. The chapter authors also request help with international ethical issues and improving ethics education within their respective countries. More collaborative research and dialogue is needed to better understand how moral distress is similar to, or differs in unique ways within international communities; and, how the nursing, medical, and bioethical communities can work together toward supportive strategies that mitigate the moral distress that imperils the health care workforce.

8.2   European Perspectives on Moral Distress

An LievrouwBo Van den Bulcke, Dominique Benoit, and Ruth Piers

8.2.1   Moral Distress Studies in Europe

Over the last 5 years, moral distress publications have increased, also in Europe, but the topic especially draws attention in North America, mirroring a broader interest in ethical practice and professional integrity in American literature [1]. Most European research is done in Western-European countries, whereas research in

Central and Eastern European countries [2] remains sparse. Insights in countries with different histories or healthcare systems are therefore overlooked [3].

Quantitative methods clearly dominate European moral distress research [1]. When different European countries are included in one study, sample sizes are small and there is often a lack of proportional representation of respondents per country and a lack of sample heterogeneity [4]. Moreover, many studies use the Moral Distress Scale [5], featuring situations that are closely linked to a North American context [6]. Review studies, often based on major databases, risk not having covered those relevant European studies that were not published in traditional outlets or were not written in English [1]. In addition, the use of various theoretical frameworks and unclear conceptualization lead to highly fragmented research material, making it more difficult to compare and integrate [7].

End-of-life care draws a great deal of attention in European moral distress research. An important common cause of moral distress in European research is the perception of healthcare providers that excessive medical treatment is provided. In a non-European country such as Iran, the maintenance of hope is considered very important in the care of the dying [8], whereas in many European studies, moral distress is especially linked to perceived overtreatment, not under-treatment [2, 4, 9–14]. Especially in an intensive care unit (ICU) setting, ethical decision-making, including decisions regarding end-of-life care, is a part of the daily tasks. Decisions are very complex, both in medical and non-medical ways [12]. Thus, because of the ethical and technological complexity in ICU care, our chapter will examine and summarize the experiences of moral distress in European-ICU settings [4, 9, 10, 14].

When dealing with ethical issues concerning the end of life, there seems no such thing as 'the' European culture. Different European countries have different opinions about when care is to be considered appropriate, and the type of decision-making varies among countries [15], also regarding decisions on life-sustaining treatment [16]. Although relevant domestic law in the various countries shows no significant differences and is based on the same international conventions [17], cultural differences do exist. In a European NICU (neonatology intensive care unit) end-of-life care study, culture-related factors (e.g., social values, cultural beliefs) seemed more relevant in end-of-life decision making than characteristics of individual physicians or units [15].

Moral distress in Europe, in line with non-European data, is linked to a reduced level of feeling satisfied at work and to an individual's sense of powerlessness [4, 9, 18]. Other moral distress triggers are the inability to influence medical decisions related to patients' level of pain and suffering [14], but also having to comply with families' wishes for patient care even though the patient disagrees [9], working with staffing levels perceived as unsafe [2, 6, 9, 11] or working with perceived incompetent colleagues [6].

8.2.2 From Individual Caretaker Viewpoint to Team Perspective

In Europe, moral distress is mostly studied in *nurses* [1], assuming that they are unable to act upon their beliefs and don't have the power to make final decisions [17]. Lower autonomy in European ICU nurses is associated with increased frequency and intensity of moral distress and lower levels of nurse–physician collaboration [4]. In the ETHICUS Study (sponsored by the Ethics Section of the European Society of Intensive Care Medicine), physicians report that nurses are involved in 95.8% of end-of-life decisions in Northern European ICUs [19], in accordance with national guidelines. But, this is in contrast to other studies that report nurses' feelings of not being involved in a satisfactory way on collaborative decisions related to withholding and withdrawing therapy in the ICU [20]. European nurses usually have higher levels of moral distress than physicians [11]; and, often feel unable to change the plan of care for their patients [10] and perceive limited control over unit operations and management issues [4]. Therefore they can be prone to behave in conformist ways, following orders, and putting aside their values and principles by capitulating to the decisions made by others [7]. Furthermore, they may consider leaving their position more frequently when experiencing moral distress [21].

Studies about moral distress in *physicians* remain scarce [17]. European physicians often experience moral dilemmas, but very few have access to support for resolving them [22]. They struggle with prognostic uncertainty and perceived pressure from referring physicians, and therefore sometimes decide to "wait and see," leading to continued disproportionate care, which can be perceived as avoiding having to take a decision [10]. In a Flemish qualitative study, physicians reported more rational ways of coping with moral distress than their nursing colleagues; and, junior physicians didn't always feel confident enough to put forward their own moral beliefs, possibly for fear of reprisal [23].

European data *combining nurses' and physicians' points of view* are even sparser. Few studies show the importance of an interdisciplinary viewpoint that encompasses a team context [1, 4, 6, 9, 11, 18], organizational culture [2, 9], and training and leadership [7]. Nurses, doctors, and other staff members do not always agree on what constitutes a moral issue. This could be due not only to different moral opinions but also to differences in knowledge and access to diverging facts about the particular situation [17]. This suggests that interprofessional imbalances might arise when having to address morally complex situations, when leaving no space for acknowledging and understanding differences. Each caretakers' viewpoint should be taken into account when addressing moral issues in healthcare settings.

The European and US Appropricus (Appropriateness of care in ICUs) [24, 25] studies, discussed in the next paragraphs, specifically aimed at integrating this team perspective, analyzing the perception of inappropriate care by health care providers in relation to individual patients' situations.

8.2.3   Results of the APPROPRICUS Study

The main objectives of the cross-sectional APPROPRICUS study published in 2011 were (1) to determine the prevalence of perceived inappropriate care among European nurses and physicians, (2) to describe the patient-related situations associated with perceived inappropriateness of care, and (3) to explore the level of agreement among clinicians concerning perceived situations of inappropriateness of care. In this study, inappropriateness of care was defined as a specific patient-care situation in which the clinician acts in a manner contrary to his or her personal and/ or professional beliefs, hence embodying a specific type of moral distress. The researchers hypothesized that perceived inappropriateness of care is not only associated with patient care situations, but also with socio-demographic and work-related characteristics and intention to leave one's job. The study was a single-day cross-sectional study among 1651 health care providers in 82 ICUs.

Although there was a wide variation in prevalence across ICUs and across countries, 27% of the European physicians and nurses declared that they had to treat at least one patient who received inappropriate care on the day of the study. Remarkably, physicians reported a higher rate of perceived inappropriate care than nurses did, contradicting other research findings suggesting that nurses usually have higher moral distress levels [11]. The two most common reported reasons were disproportion between the amount of care given and the expected prognosis (65%), of which too much (89%) or too little (11%) care was the concern, and the feeling that other patients would benefit more from the ICU care than the present patient (38%). Clinicians reported that they were distressed by the perception of inappropriate care. The study revealed five protective factors: decisions about symptom control shared by nurses and physicians; involvement of nurses in end-of-life decisions; good collaboration between nurses and physicians; work autonomy and perceived lower workload.

A cross-sectional study in California conducted in 2013 within 56 ICUs yielded similar results [24]. The prevalence of perceptions of inappropriate care was even higher among the respondents. Thirty-eight percent of 1,169 health care providers (51.1% of physicians and 35.8% of nurses) identified at least one patient as receiving inappropriate treatment. Respondents most commonly reported that the amount of treatment provided was disproportionate to the patient's expected prognosis or wishes (76%) and that treatment was "too much" in 93% of cases. Factors associated with perceived inappropriateness of treatment were the belief that death in their ICU is seen as a failure, profession (doctors more than nurses), lack of collaboration between doctors and nurses, intent to leave their job, and the perceived responsibility to control health-care costs. Health care providers supported formal communication training and mandatory family meetings as potential solutions to reduce the provision of inappropriate treatment.

The APPROPRICUS study in Europe and the US both conclude that perceived inappropriateness of care will always be part of health care. Dealing with uncertainty (even with the integration of palliative care) can lead to morally distressing situations especially when organizational systems are lacking in supportive management and other needed clinician resources. The successful management of moral distress can be an opportunity for transformation and growth of a team. Attention to moral distress is relevant to the development of a strong ethical climate in a workplace and the prevention of the negative outcomes associated with moral distress. There is a potential learning for the individual as for the team that may result from the experience of moral distress [26]. Expressing a perception of excessive care to colleagues requires a safe climate in which clinicians are empowered to speak up and in which they feel that their opinions are valued and subsequently integrated into the decision-making process [27, 28]. Moreover, implementing active communication regarding end-of-life care in the ICU seems associated with lower health care workers' suffering [29] and lower long-term anxiety in relatives [30].

Team reflection could be regarded as a key factor in predicting effective teamwork [31]. Studies show that self-reflection is done too infrequently within health care organizations [32]. In fact, it seems essential that the entire ICU team receives structured opportunities to safely work through ethical dilemmas and conflicts [20]. Debriefings and ethical discussions become critical survival tools, allowing clinicians to recognize and resolve distress. Proper support and training are indispensable, and there is a need to design and evaluate effective ethics competence programs [33].

Physicians should use their leadership function as a role-model for the team to enhance self-reflection and stimulate other team members to build an open communication culture, yet they may also need training to serve in this leadership role [28]. Interventions that improve mutual understanding, communication, and cooperation among various healthcare disciplines, thus also facilitating leadership, may help alleviate ethical problems and enhance the quality of patient care [12, 31–33]. There is a critical need for education to improve ethical understanding, ethical skills, and communication. Clinicians also need morally sensitive support mechanisms within their organizational workplaces, including nurse-physician mentorship models that aid collaborative practice. This might include time for individual engagement in critical self-reflection and structured interdisciplinary dialogue such as "death rounds" (structured end-of-life discussions among team members) [34].

8.2.4 Conclusion

Moral distress is a challenging phenomenon in European ICU Care, although there are varying results in different European countries and the small number of studies, small samples, and a lack of sample heterogeneity do not allow generalization. Qualitative studies rooted in a European perspective remain scarce, yet qualitative designs can shed more in-depth light on the complex process of ethical practice [7].

Different European countries have different views on when care becomes inappropriate consistent with their sociocultural and political contexts. However, an important common ground for moral distress is the perception of overtreatment, especially in end-of-life care. Perceived inappropriateness of care will always be part of health care. Early discussions of patients' wishes with regard to treatment options is important in preventing overtreatment [12, 35]. Despite the fact that it is not always possible to anticipate a patient's life expectancy, health professionals should aim at constructing care trajectories that address patients' different illness experiences and needs, helping patients, families, and professionals cope with the situation [35]. A safe climate and a culture of self-reflection and open communication doesn't come naturally in European ICU settings. Physicians should use their leadership to help create such a work environment and should facilitate ethical debate. To encourage autonomy, nurses should be more closely involved in unit-level decisions and organizational goals, and a critical engagement in ethical dialogue should be supported [4]. Moral distress can thus be both a challenge and an opportunity; it may potentially lead to increased introspection and team-reflection, mutual understanding and communication among the disciplines, and individual and team growth, with the ultimate goal of generating better patient and family care.

Suggested Reading

Lamiani G, Borghi L, Argentero P. When healthcare professionals cannot do the right thing: a systematic review of moral distress and its correlates. J Health Psychol. 2017;22(1):51–67.

Atabay G, Cangarli BG, Penbek S. Impact of ethical climate on moral distress revisited: multidimensional view. Nurs Ethics. 2015;22(1):103–16.

Hurst SA, Perrier A, Pegoraro R, Reiter-Theil S, Forde R, Slowther AM, et al. Ethical difficulties in clinical practice: experiences of European doctors. J Med Ethics. 2007;33(1):51–7.

Papathanassoglou ED, Karanikola MN, Kalafati M, Giannakopoulou M, Lemonidou C, Albarran JW. Professional autonomy, collaboration with physicians, and moral distress among European intensive care nurses. Am J Crit Care. 2012;21(2):e41–52.

Corley MC, Elswick RK, Gorman M, Clor T. Development and evaluation of a moral distress scale. J Adv Nurs. 2001;33(2):250–6.

Silen M, Svantesson M, Kjellstrom S, Sidenvall B, Christensson L. Moral distress and ethical climate in a Swedish nursing context: perceptions and instrument usability. J Clin Nurs. 2011;20(23-24):3483–93.

Goethals S, Gastmans C, de Casterle BD. Nurses' ethical reasoning and behaviour: a literature review. Int J Nurs Stud. 2010;47(5):635–50.

Shorideh FA, Ashktorab T, Yaghmaei F. Iranian intensive care unit nurses' moral distress: a content analysis. Nurs Ethics. 2012;19(4):464–78.

de Veer AJ, Francke AL, Struijs A, Willems DL. Determinants of moral distress in daily nursing practice: a cross sectional correlational questionnaire survey. Int J Nurs Stud. 2013;50(1):100–8.

Piers RD, Azoulay E, Ricou B, DeKeyser Ganz F, Max A, Michalsen A, et al. Inappropriate care in European ICUs: confronting views from nurses and junior and senior physicians. Chest. 2014;146(2):267–75.

de Boer J, van Rosmalen J, Bakker A, van Dijk M. Appropriateness of care and moral distress among neonatal intensive care unit staff: repeated measures. Nurs Crit Care. 2015;21:19–27.

Oerlemans AJ, van Sluisveld N, van Leeuwen ES, Wollersheim H, Dekkers WJ, Zegers M. Ethical problems in intensive care unit admission and discharge decisions: a qualitative study among physicians and nurses in the Netherlands. BMC Med Ethics. 2015;16:9.

Sannino P, Gianni ML, Re LG, Lusignani M. Moral distress in the neonatal intensive care unit: an Italian study. J Perinatol. 2015;35(3):214–7.

Lusignani M, Gianni ML, Re LG, Buffon ML. Moral distress among nurses in medical, surgical and intensive-care units. J Nurs Manag. 2016;25(6):477–85.

Cuttini M, Nadai M, Kaminski M, Hansen G, de Leeuw R, Lenoir S, et al. End-of-life decisions in neonatal intensive care: physicians' self-reported practices in seven European countries. EURONIC Study Group. Lancet. 2000;355(9221):2112–8.

Sprung CL, Cohen SL, Sjokvist P, Baras M, Bulow HH, Hovilehto S, et al. End-of-life practices in European intensive care units: the Ethicus Study. JAMA. 2003;290(6):790–7.

Kalvemark S, Hoglund AT, Hansson MG, Westerholm P, Arnetz B. Living with conflicts-ethical dilemmas and moral distress in the health care system. Soc Sci Med. 2004;58(6):1075–84.

Karanikola MN, Albarran JW, Drigo E, Giannakopoulou M, Kalafati M, Mpouzika M, et al. Moral distress, autonomy and nurse-physician collaboration among intensive care unit nurses in Italy. J Nurs Manag. 2014;22(4):472–84.

Benbenishty J, Ganz FD, Lippert A, Bulow HH, Wennberg E, Henderson B, et al. Nurse involvement in end-of-life decision making: the ETHICUS Study. Intensive Care Med. 2006;32(1):129–32.

Jensen HI, Ammentorp J, Erlandsen M, Ording H. Withholding or withdrawing therapy in intensive care units: an analysis of collaboration among healthcare professionals. Intensive Care Med. 2011;37(10):1696–705.

Oh Y, Gastmans C. Moral distress experienced by nurses: a quantitative literature review. Nurs Ethics. 2015;22(1):15–31.

Forde R, Aasland OG. Moral distress among Norwegian doctors. J Med Ethics. 2008;34(7):521–5.

Lievrouw A, Vanheule S, Deveugele M, Vos M, Pattyn P, Belle V, et al. Coping with moral distress in oncology practice: nurse and physician strategies. Oncol Nurs Forum. 2016;43(4):505–12.

Piers RD, Azoulay E, Ricou B, Dekeyser Ganz F, Decruyenaere J, Max A, et al. Perceptions of appropriateness of care among European and Israeli intensive care unit nurses and physicians. JAMA. 2011;306(24):2694–703.

Anstey MH, Adams JL, McGlynn EA. Perceptions of the appropriateness of care in California adult intensive care units. Crit Care. 2015;19:51.

Musto LC, Rodney PA, Vanderheide R. Toward interventions to address moral distress: navigating structure and agency. Nurs Ethics. 2015;22(1):91–102.

Reader TW, Flin R, Mearns K, Cuthbertson BH. Interdisciplinary communication in the intensive care unit. Br J Anaesth. 2007;98(3):347–52.

Van den Bulcke B, Vyt A, Vanheule S, Hoste E, Decruyenaere J, Benoit D. The perceived quality of interprofessional teamwork in an intensive care unit: a single centre intervention study. J Interprof Care. 2016;30(3):301–8.

Quenot JP, Rigaud JP, Prin S, Barbar S, Pavon A, Hamet M, et al. Suffering among carers working in critical care can be reduced by an intensive communication strategy on end-of-life practices. Intensive Care Med. 2012;38(1):55–61.

Hartog CS, Schwarzkopf D, Riedemann NC, Pfeifer R, Guenther A, Egerland K, et al. End-of-life care in the intensive care unit: a patient-based questionnaire of intensive care unit staff perception and relatives' psychological response. Palliat Med. 2015;29(4):336–45.

Bruce CR, Miller SM, Zimmerman JL. A qualitative study exploring moral distress in the ICU team: the importance of unit functionality and intrateam dynamics. Crit Care Med. 2015;43(4):823–31.

Piquette D, Reeves S, LeBlanc VR. Stressful intensive care unit medical crises: how individual responses impact on team performance. Crit Care Med. 2009;37(4):1251–5.

Kalvemark Sporrong S, Arnetz B, Hansson MG, Westerholm P, Hoglund AT. Developing ethical competence in health care organizations. Nurs Ethics. 2007;14(6):825–37.

Hough CL, Hudson LD, Salud A, Lahey T, Curtis JR. Death rounds: end-of-life discussions among medical residents in the intensive care unit. J Crit Care. 2005;20(1):20–5.

Murray SA, Kendall M, Boyd K, Sheikh A. Illness trajectories and palliative care. BMJ. 2005;330(7498):1007–11.

8.3   Speaking Truth to Power

Georgina Morley

Like many nurses, I became interested in moral distress because of my own experiences. These experiences were similar to the "classic" moral distress narratives that can be found in the literature; a young nurse, unheard, lacking confidence and experience, feeling powerless and constrained to act. Yet there were also significant differences. I was not morally certain. I didn't feel morally distressed because, as per Jameton [36], I *knew* the right thing to do but was prevented by institutional constraints. Instead, my moral distress stemmed from a sense of deep unease that we simply weren't managing morally challenging situations. We were failing our patients through our inaction and inability. In this narrative I am going to discuss some of the constraints on my moral agency that I believe contributed to my experiences of moral distress: a lack of clinical ethics education; an environment which lacked moral spaces and moral community; and feeling marginalized and disempowered.

I went into nursing as a philosophy graduate. My theoretical knowledge of ethics made me acutely aware of the very real ethical challenges I faced, for example: how to manage a mobile adolescent who had a craniectomy and unsteady gait but refused to wear his protective helmet; whether or not to tell a patient's partner of their HIV status; how to discharge a homeless patient back onto the streets. All of these challenges, and more, caused me to feel morally distressed—not because I felt I knew, with any degree of certainty, the "right thing" in each of these scenarios but because I didn't know how to navigate them. What my philosophy and nursing education had failed to prepare me for was how to think about and manage ethical problems in real life. The situations I was facing on a regular basis were morally fraught and no one was addressing them, or if they were, they weren't communicating to the frontline staff. Hospital life appeared to go on unchanged. I would voice my concerns to the doctors: "Should we give the patient something to calm them down so they don't hurt themselves?" Or when the patient started to become physically abusive to the nursing staff: "Should we give the patient something now?" I wasn't assertive in my questioning, I was unsure; I didn't know what the right thing was. On reflection, I'm not sure that they did either.

I came across research exploring moral distress in the United States (US) and I was amazed that it was being researched so extensively. What surprised me even more was the lack of published research from the United Kingdom (UK) exploring the concept. I wonder now whether this lack of awareness in the UK stems from a lack of engagement with clinical ethics more generally. Anecdotally, my own experiences suggest there is very little ethics within the nursing curriculum and we lag behind the US in that respect. In my own research, in which I explore nurses' experiences of moral distress, the nurses I have interviewed to date have expressed the feeling that their ethics education was also lacking. They seem underprepared to deal with ethical issues they encounter in their practice. Based on my general observations, I suggest that some nurses' understanding of ethics is so limited that they only recognize the "big" ethical issues, such as the withdrawal of life-sustaining treatments and fail to recognize the "microethics" that permeate their every encounter with patients [37].

In the UK, The Nursing and Midwifery Council (NMC) are responsible for setting the standards by which nurses are expected to practice (The Code) and set the curriculum for undergraduate nurses. The Code sets out statements that represent good nursing practice: to prioritize people, practice effectively, preserve safety, and promote professionalism and trust [38]. Undoubtedly these are standards to which we must aspire in our everyday nursing practice, but simply codifying one's general obligations is problematic. The simple codification of general professional values is not sufficiently action-guiding and The Code does not provide practical advice about navigating everyday clinical ethical dilemmas. The Code recommends that I both prioritize people and preserve safety; so where does this leave me with regard to the craniectomy patient who wants to wander the halls of the hospital and the medics who do not want to sedate him? Do I prioritize patient choice by allowing the patient to mobilize and thereby allow him to be both a risk to himself and myself given his volatile nature, or do I insist on preserving safety and beg the medics to sedate him? I'm certainly not promoting professionalism or trust to the visitors now entering the unit, who are watching me inexpertly attempt to convince this young patient that he ought to wear his protective helmet whilst three other patient call bells are going off.

If we turn to the NMC standards for competence regarding pre-registration and registered nurses, the NMC state nurses must:

> '…act with professionalism and integrity, and work within agreed professional, ethical and legal frameworks and processes to maintain and improve standards' (p. 4; p. 5) [39, 40].

Programme providers are responsible for ensuring "professional codes, ethics, law and humanities" are included in their programme content and underpin practice (p. 73) [40] , meaning that the degree to which ethics will feature in nurse training is greatly dependent upon higher education institutions (HEIs). It is up to HEIs to supplement future professionals' normative commitments with the skills to navigate clinical ethical problems. Without specific training, frameworks, or tools for moral decision-making, greater emphasis is placed on individual's own moral intuitions in order to fulfill their professional obligations. Unfortunately, moral intuitions are often either in conflict or contradictory which makes them unreliable. Furthermore, Hamric et al. [41] argue that the requirement to satisfy professional obligations has increasingly become reliant upon the notion of "moral courage," and that the invocation of moral courage has become excessive. Drawing on the work of philosopher Lisa Tessman [42], they argue that calling for clinicians to be courageous represents the daily oppression of clinical practice [41].

Rather than calling on clinicians to be courageous in order to reduce moral distress, we need to place greater emphasis on ethics education and the creation of moral spaces where professionals can discuss, reflect, and explore ethical issues together [43]. Building on work by Margaret Urban Walker [44], and Hamric et al. [43] argue that institutions in the US must go beyond the standard creation of clinical ethics committees (CECs) in order to satisfy The Joint Commission's requirement for processes to deal with ethical issues. They argue that ethics resources must be knowledgeable, known, available, and sanctioned in order to truly carve safe spaces for moral deliberation [43]. Indeed, evidence suggests that moral distress is reduced when there is a greater sense of moral community and nurses feel they have supportive relationships, are able to express their moral uncertainty, and can manage conflict [45].

I have highlighted elsewhere how the UK has been slow to adopt CECs, partly because they are not mandated, and how few institutions have the resources available for ethics consultants and consultations [46]. Bioethicists and clinicians must continue to advocate for further integration of ethics resources, not just within the institution but by the bedside, where the majority of nurses spend their time; and for the need to sanction ethics resources if we want them to thrive. Echoing a critique originally made by Warren [47], Fitzpatrick et al. [48] argue that bioethics must not be concerned only with "crisis" issues but must also focus on personal or "housekeeping issues" (p. 64). Whilst many bioethicists are understandably distracted by the emergence of new biotechnologies, this should not be at the expense of everyday clinical moral deliberations: the microethics of practice.

Feminists have long argued that caring professions and "dependency work[ers]" (p. 444) [49] fail to have their needs recognized within society and that this needs to change [47]. Rather than only exploring crisis issues, I urge ethicists to also explore the everyday clinical ethical issues and the knock-on effects on clinicians. It is little wonder that nurses have been so fascinated by the phenomenon of moral distress since it captures so well their everyday experiences of marginalization and disempowerment. Yet in recent years it has become more fully recognized and acknowledged that moral distress does not only affect nurses but also all members of the multi-disciplinary team [50]. With this, I think we nurses are losing possession of our precious moral distress as a driver for change. It will no longer capture our peculiar experiences but the experiences of the whole healthcare team. In many ways this is good. Perhaps if moral distress is perceived to only be a phenomenon within nursing it will never garner enough attention. Perhaps with greater recognition of the effects of moral distress on a wider clinical group, institutions will be forced to examine their ethics resources, sanction CECs, and create moral spaces. Hopefully, over time the healthcare hierarchy will be transformed into a moral community [45] because I agree with Hamric et al. [41] that the requirement of courage does itself suggest oppression and reliance on courage, which is arguably supererogatory, and is not sustainable.

Yet the loss of moral distress for nurses does risk the loss of a weapon from our arsenal since it has been a driver for change in the US and I hope that the same will apply in the UK. So the question remains therefore how we can both expand our moral distress research whilst also ensuring that the nursing voice is still heard? It is likely that this requires academic activism to advocate for our clinical colleagues. I take inspiration from feminist ethicists and philosophers who argue that morality is not a unified knowledge but is a "social medium in which people employ their shared moral understandings to carry out, contest or negotiate responsibilities" (p. 45) (Lindemann [51] on Walker [52]). Feminist ethics grew out of a movement against traditional moral philosophy in which the moral agent was viewed as an autonomous and rational actor, rationally deliberating from universal, abstract principles about the "right" thing to do and "unburdened by the non-ideal constraints of luck (moral and otherwise), circumstance and capability" [53]. The feminist view positions ethics in the social world, as part of particular historical and cultural locations, and addresses the dominance of some voices over others [54]. This view of morality stands nurses in very good stead since it prioritizes marginalized voices and embraces the holistic knowledge that nurses often pride themselves on. Being by the bedside means that nurses often know the most about patients and families, their social circumstances and dynamics, which, on this view, are crucial for moral deliberation. Academic activists and clinicians need to speak truth to power and argue for better ethics education with resources to supplement recognized professional obligations; the need for greater moral space and the integration of marginalized voices into moral deliberation. Clinical nurses need to make clear the specialist holistic knowledge that they can contribute to these discussions and unfortunately this may require a little more courage yet.

Suggested Readings

Jameton A. Nursing practice: the ethical issues. New Jersey: Prentice Hall; 1984.

Truog RD, Brown SD, Browning D, Hundert EM, Rider EA, Bell SK, et al. Microethics: the ethics of everyday clinical practice. Hastings Cent Rep. 2015;45(1):11–7.

Nursing & Midwifery Council N. The Code: professional standards of practice and behaviour for nurses and midwives: Nursing & Midwifery Council N. 2015.

Nursing & Midwifery Council N. Standards for pre-registration nursing education. London: Nursing & Midwifery Council, NMC. 2010. p. 1–152.

Nursing & Midwifery Council N. Standards for competence for registered nurses. London: Nursing & Midiwfery Council, NMC. 2014. p. 1–21.

Hamric AB, Arras JD, Mohrmann ME. Must we be courageous? Hast Cent Rep. 2015;45(3):33–40.

Tessman L. Burdened virtues: virtue ethics for libertory struggles. New York: Oxford Scholarship Online; 2005.

Hamric A, Wocial LD. Institutional ethics resources: creating moral spaces. Hast Cent Rep. 2016;46(Suppl 1):S22–7.

Walker MU. Keeping moral space open: new images of ethics consulting. Hast Cent Rep. 1993;23(2):33–40.

Traudt T, Liaschenko J, Peden-McApline C. Moral agency, moral imagination, and moral community: antidotes to moral distress. J Clin Ethics. 2016;27(3):201–13.

Morley G. Efficacy of the nurse ethicist in reducing moral distress: what can the NHS learn from the USA? Part 2. Br J Nurs. 2016;25(3):156–61.

Warren VL. Feminist directions in medical ethics. Hypatia. 1989;4(2):73–87.

Fitzpatrick P, Scully JL. Theory in feminist bioethics. In: Scully JL, Baldwin-Ragaven LE, Fitzpatrick P, editors. Feminist bioethics at the center, on the margins. Baltimore: Johns Hopkins University Press; 2010. p. 61–9.

Kittay EF, Jennings B, Wasunna AA. Dependency, difference and the global ethic of longterm care. J Polit Philos. 2005;13(4):443–69.

Whitehead PB, Herbertson RK, Hamric AB, Epstein EG, Fisher JM. Moral distress among healthcare professionals: report of an institution-wide survey. J Nurs Scholarsh. 2014;47(2):117–25. https://doi.org/10.1111/jnu.12115.

Lindemann H. Speaking truth to power. Hast Cent Rep. 2010;40(1):44–5. https://doi.org/10.1353/hcr.0.0215.

Walker MU. Moral understandings: a feminist study in Ethics. 2nd ed. New York: Oxford University Press; 2007.

Gotlib A. Feminist ethics and narrative ethics. Internet encyclopedia of philosophy. 2014. http://www.iep.utm.edu/fem-e-n/. Accessed 11 Nov 2016.

Walker MU. Introduction: Groningen naturalism in bioethics. In: Lindemann H, Verkerk M, Walker MU, editors. Naturalized bioethics: toward responsible knowing and practice. New York: Cambridge University Press; 2009. p. 1–20.

8.4 Moral Distress in the Provision of Health Care in Tanzania: Developing World Perspective

Renatha Joseph and Baraka Morris

"Moral distress is experienced by many health providers in the developing world and sometimes it makes me feel so uneasy. In 2014, I was working at Shinyanga regional hospital as a medical specialist in the Department of Pediatrics. One of the major challenges of working as a Pediatrician in the developing world is that children die of anemia due to their low blood hemoglobin and limited resources. One morning I reported to the hospital to attend to my pediatric patient and I realized that I had six patients who needed an urgent blood transfusion. The laboratory technician reported that there was no blood available and to make matters worse one of the mothers had lost her five children due to anemia. And, in front of me was the sixth child; and, she had only one child left at home. The probability of saving the child was too small. Fortunately the child was saved because relatives donated blood. The blood, however, was screened at the local hospital, but due to limited resources the blood could not be sent to the zonal blood bank for DNA Polymerase Chain Reaction (PCR) for HIV testing. At the time, I felt useless that I could not save the patient's life which was my primary obligation at that moment. What was I to do? Should I let the child die of anemia or give blood that was not screened enough to be sure that it was safe? The moment was so disappointing and painful, as a mother myself, pediatrician and children's advocate."… Tanzanian Pediatric Physician

Moral distress in health care is experienced when healthcare personnel (e.g. doctors and nurses) cannot meet their professional goals and clinical standards of care due to institutional or other types of constraints [36]. The ultimate goal of medicine is to save lives and alleviate suffering. In the same way, the Tanzanian Code of Ethics requires nurses to "respect humankind and life, obtain consent for care, maintain professional competence, exercise trustworthiness and fairness, protect confidential information, and take responsibility for their actions." (p. 5) [55]. But as evidenced in the above scenario, many children suffer due to limited resources in global communities, as do the providers who care for them. Elpern et al. [56] define moral distress as "painful feelings and/or psychological disequilibrium that occurs in situations in which the ethically right course of action is known but cannot be acted upon. As a result, persons in moral distress act in a manner contrary to their personal and professional values (p. 523)." The pediatrician in the case exemplar felt "useless" and "helpless" to do anything, yet was professionally and morally required to "do something" to help the child during the crisis.

In the developing world, health care workers can become morally distressed as a result of the various ethical dilemmas they encounter in their daily work life. This mostly centers on the allocation of scarce resources and includes everyday decisions on who to save and who to let die due to the lack of health resources. For children and others who need blood, this might mean the lack of a blood transfusion. For other patients, this could mean waiting until a medication becomes available. Depending on the severity of the patient's condition and clinical diagnosis, patients may get better for self-limiting conditions, but they also might suffer from disability or even death. In addition, there are no clear ethical guidelines on priority setting when all patients cannot receive immediate care. Informed consent is also not clearly understood in the developing world and particularly its importance in the patient–provider relationship. It is further encumbered by limited time to engage in the informed consent process. Healthcare workers are understaffed and underpaid for the work that is required of them, personal values often clash with professional values (especially, for example, related to family planning methods, abortion, and letting patients die), competing responsibilities place pressures on nurses when they receive orders from different authorities; and finally, ethically "right" actions from the perspective of physicians and nurses are not always legally "right" within the country. All of these ethical issues lead to a morally distressed healthcare workforce in Tanzania.

The severe scarcity of material and human resources in the developing world places an undue burden on health care workers [57], potentially leading to morally distressing situations and a loss of dedicated professionals [58]. In 2010, the shortage of skilled personnel in Tanzania, for example, was estimated to be 62–68% of what was needed (Medical Association of Tanzania) (MAT) [59]. The shortages, however, are not equally distributed with urban–rural imbalances. In 2012, the shortage was 65% with a doctor-to-patient ratio of 1:25,000 [60, 61].

According to Hellsten [62], health care workers might also need to perform duties above and beyond their level of training because of the scarcity of providers within particular regions of the developing country. Moreover, material resources for advanced treatment for certain healthcare conditions might be available at expensive private hospitals but not available at public or governmental institutions. Poor patients cannot afford to buy medicines or pay for expensive treatments when the average income in Tanzania is extremely low. The multidimensional poverty index reports that 31.9% of the population live in extreme poverty with a Gross National Income (GNI) per capita of $1,702 and 9.7% of Tanzanians live below $1 per day [63]. As a clinician, you start wondering what more could be done to make services available for all patients. Distress ensues because you are deciding to give suboptimal care to patients who cannot afford to buy the medicine or services that would necessarily help them. Despite the fact that health care services are free to certain groups of patients according to Tanzanian national policy (including pregnant women, children and the elderly), the availability and accessibility of medical treatment to these groups is often limited. This situation is comparable to the moral distress reported by an internal medicine resident in a case written by Hamric, Davis, and Childress [64]. In this case, the medical resident was distressed thinking of a patient's situation because the patient was unable to pay and had no health insurance. In Tanzania, just 30% of the population has health insurance among which 7.3% of these individuals receive this coverage from the National Health Insurance Funds (NHIF). However, the majority of citizens have what is known as care through a Community Health Fund (CHF) [65] which provides for a few health care services within the same district where an individual joined the CHF. The service package is at the discretion of the specific district and includes mostly primary care services such as clinician consultations, uncomplicated deliveries, and simple laboratory services (e.g., checking blood for the presence of the malaria parasite, blood sugar testing and urinalysis) [60].

Health care services given to each patient in Tanzania have to be clearly explained for the patient to understand their care needs, such as prescriptions for a particular condition and how to use the given medication. But, doctors need to have enough time to give focused information to the competent patient or surrogate decision maker [66]. Unfortunately a single doctor can become exhausted and overwhelmed by the patient load, sometimes averaging 29 patients in a single day [67]; and therefore, there is not enough time to fully explain procedures, tests, or recommended treatments to patients. And, there is also minimal time for question and answer. This creates ethically challenging situations because the doctor knows that he or she is supposed to take informed consent seriously but there is not enough time to meet the needs of every patient. Time constraints in clinical care are not the only problem for practicing patient centered care; the level of education and medical literacy of patients during the medical encounter is also challenging, especially since literacy is low in the developing world. In 2012, the literacy level in Tanzania was approximately 71.8% [68]. The combination of time constraints and low patient literacy hamper what clinicians can accomplish and greatly contribute to moral distress of the health care workforce.

In the developing world there is a scarcity of important medical equipment and supplies. This scarcity of medical supplies raises many ethical concerns [69]. Every day health care workers have to ration treatment and service delivery for their patients. They have to decide what they can and cannot do. For example, in some of the district hospitals there is only one oxygen tank available in the institution, and there might be 4 or 5 patients who need oxygen on that particular day. Without it, however, these patients will not survive. The lack of essential equipment and supplies includes the need for blood products when blood transfusions are required as reflected in the case exemplar. Donated blood is kept in the blood bank for later use but it first must be taken to a zonal blood bank for screening of infectious diseases like HIV, hepatitis, and other maladies. When there is no centrally screened blood, doctors have to decide to screen the blood product at their local hospital for donation. But, what if something is missed due to the lack of proper screening, and how does one balance benefits and risks to patients? Here, ethical dilemmas arise because the doctor has to determine whether to transfuse the patient with a questionable blood product wondering whether the patient might or might not become infected at a later point in time. Or, let the patient die of heart failure secondary to severe anemia.

Nurses in Tanzania are also under tremendous pressure because they receive instructions, orders, and requests from authority figures that include physicians but also patients. And, sometimes pressure comes from patients' relatives or family members. More than a quarter of all health care workers in Tanzania are nurses (27%); and, doctors represent 1.7% of this population [70, 71]. The ratio of nurses to the population in Tanzania is 0.39:10,000 [71, 72]. This number is low compared to what is needed, and typically one registered nurse is supposed to attend 8 patients with the assistance of two enrolled nurses (enrolled nurses are those who attended nursing school for 3 years and have a certificate) [73]. In actual practice there are two registered nurses and four enrolled nurses in the ward with 40 to 60 patients. They have to stretch the resources they have to be able to provide health care on a daily basis. Nurses have to write a nursing report for each patient every day and every shift; and, this must be presented to the nurse in-charge during hand offs from one duty staff to another. Unfortunately, distress occurs for nurses when they feel as though they cannot provide care for all of their patients, have little time for documentation, and try to document as needed in the reality of insufficient staff and other resources.

Issues of nurse–physician collaboration are also important in the developing world and if not handled well, distressing events become exaggerated. In current qualitative work examining nurse–physician collaboration in African countries, Tanzanian nurses report that some doctors are harsh and arrogant; and, some young doctors think nurses do not trust their abilities to provide proper patient care [74]. Lack of trust and professional collaboration between nurses and physicians creates moral distress and this distress is heightened by the lack of an inner motivation to work collaboratively together. We also found that health care providers are less motivated if their salaries are suboptimal. To some degree, moral distress can be mitigated by developing ethics based competencies in Tanzanian health care workers through ethics educational requirements and other in-house resources. A new Department of Bioethics and Health Professionalism was established in 2016 at Muhimbili University in Dar Es Salaam, preparing students to address ethical concerns in the country [75, 76]. The department offers courses on Bioethics and Health Professionalism (including clinical and research ethics) to both undergraduate and postgraduate students, preparing future health care providers to work in an ethical manner and helping to reduce misunderstandings in patient care. Ultimately, these programs aim to improve patient outcomes and the health of all Tanzanian citizens.

To improve patient outcomes and reduce moral distress in the developing world among health care workers, there is an urgent need to improve working conditions. Institutions and the government should outline clear job descriptions and policies that support nurses and physicians currently practicing in Tanzania. For example, what does nursing care entail when resources are limited and one cannot meet the needs of all their patients? How do physicians and nurses meet their moral and professional commitments to their patients and families when they are morally distressed by their working conditions? How do Tanzanian nurses and physicians collaborate more effectively to meet the needs of their patients? More dialogue is needed on the types of institutional committees (e.g., hospital ethics committees) that will support clinicians who are facing ethical dilemmas in their service provision. Distress management is one way to help train clinicians who already suffer from moral distress by providing them with strategies to resolve various ethical issues that they encounter in their daily work. There should also be a system in place that supports these nurses and physicians to reach their professional developmental goals through educational scholarships and other types of mechanisms that address the immediate needs of the host country. Supporting clinicians in this way may lead to retention of a qualified workforce that can assist the government and other institutional agencies in solving the complex health conditions of the citizens within Tanzania, leading to greater satisfaction for patients and families and the clinicians who care for them.

Suggested Readings

Jameton A. Nursing practice: the ethical issues. New Jersey: Prentice Hall; 1984.

Tanzanian and Midwifery Council. Code of Ethics and Professional Conduct for Nurses and Midwives in Tanzania Revised. Tanzania; 2015.

Elpern B, Covert B, Kleinpell R. Moral distress of staff nurses in a medical intensive care unit. Am J Crit Care. 2005;14(6):523–30.

Sirili N, Kiwara A, Nyongole O, Frumence G, Semakafu A, Hurtig AK. Addressing the human resource for health crisis in Tanzania: the lost in transition syndrome. Tanzan J Health Res. 2014;16(2):1–9.

Sikika, Medical Association Of Tanzania. Where are the Doctors ? - Tracking Study of Medical Doctors. 2013. http://sikika.or.tz/wp-content/uploads/2014/03/Practice-Status-of-Medical-Graduates-MD-Tracking-edited.pdf.

Medical Association of Tanzania (MAT). Proceedings of the 43rd Annual General Meeting and 45th Anniversary. 2010.

Kwesigabo G, Mwangu MA, Kakoko DC, Warriner I, Mkony CA, Killewo J, et al. Tanzania's health system and workforce crisis. J Public Health Policy. 2012;33(Suppl 1(S1)):S35–44. http://www.ncbi.nlm.nih.gov/pubmed/23254848

The United Republic of Tanzania, Ministry of Health and Social Welfare. Human resource for health strategic plan 2008–2013. Ministry of Health and Social Welfare. 2008.

Hellsten SK. Bioethics in Tanzania: legal and ethical concerns in medical care and research in relation to the HIV/AIDS epidemic. Camb Q Healthc Ethics. 2005;2005:256–67.

Economic and Social Research Foundation. Development report, 2014; Economic Transformation for Human Development. 2014. p. 1–128.

Hamric AB, Davis WS, Childress MD. Moral distress in health care professionals. Pharos Alpha Omega Alpha Honor Med Soc. 2006;69(1):16–23. http://www.ncbi.nlm.nih.gov/pubmed/16544460

Kuwawenaruwa A, Borghi J. Health insurance cover is increasing among the Tanzanian population but wealthier groups are more likely to benefit. Ifakara Health Institute. 2012. p. 1–4.

Vallely A, Lees S, Shagi C, Kasindi S, Soteli S, Kavit N, Vallely L, McCormack S, Pool R, Hayes RJ for the Microbicides Development Programme (MDP). How informed is consent in vulnerable populations? Experience using a continuous consent process during the MDP301 vaginal microbicide trial in Mwanza Tanzania. BMC Med Ethics 2010;11:10.

Manzi F, Schellenberg J, Hutton G, Wyss K, Mbuya C, Shirima K, et al. Human resources for health care delivery in Tanzania: a multifaceted problem. Hum Resour Health. 2012;10(1):3. http://www.human-resources-health.com/content/10/1/3

National Bureau of Statistics (NBS). The United Republic of Tanzania 2015 Tanzania in figures. 2016. http://www.nbs.go.tz/nbs/takwimu/references/Tanzania_in_Figures_2015.pdf.

Tibandebage BP, Kida T, Mackintosh M, Ikingura J. Understandings of ethics in maternal health care: an exploration of evidence from four districts in Tanzania. 2013. http://www.repoa.or.tz/documents/REPOA_WORKING_PAPER_13.2.pdf.

Kwesigabo G, Mwangu MA, Kakoko DC, Killewo J. Health challenges in Tanzania: context for educating health professionals. J Public Health Policy. 2012;33(Suppl 1):S23–34.

Kurowski C, Wyss K, Abdulla S, Yémadji D, Mills A. Human resources for health: requirements and availability in the context of scaling-up priority interventions in low-income countries Case studies from Tanzania and Chad. Heal Econ Financ Program. 2004. p. 96. https://assets.publishing.service.gov.uk/media/57a08c12e5274a31e0000f9a/WP01_04.pdf.

Munga MA, Maestad O. Measuring inequalities in the distribution of health workers: the case of Tanzania. Hum Resour Health. 2009;7:4.

Education MOF, Training V, Technology C, Syllabus A, Certificate FOR, In C, et al. the United Republic of Tanzania. 2009.

Ulrich CM, Muecke, M., Mann Wall, B., Hoke, L., Joseph, R., Shayo, J.E., Morris, B.M., Sabone, M., Cainelli, F., Mazonde, P., Maitshoko, M. Inter-professional practice and education: a collaborative initiative with Tanzania and Botswana: Penn in Africa. 2013.

Waddell R, Aboud M. Dartmouth/Muhas research ethics training and program development for Tanzania, [R25TW007693]. National Institutes of Health, Fogarty: Tanzania; 2011-2016.

Ringer S, Aboud M. Dartmouth/Muhas Research Ethics Training and Program Development for Tanzania. [R25TW007693]; Tanzania: National Institutes of Health, Fogarty; 2017-2022. National Institutes of Health, Fogarty; 2017-2022.

8.5   Truth Telling and Moral Distress : A Singaporean Perspective

Subadhra D. Rai and Margaret Mei Ling Soon

In December of 1997, my father was diagnosed with pancreatic cancer. He underwent surgery; however, we found out that it would be an "open and close" as it had metastasized to the liver. My father recovered from the surgery and was discharged home. But he did not know that the surgery was palliative and not curative and neither was he privy to the news that he had six months to live. We kept both news from him and our mother. As a nurse in the family, I struggled a lot of not informing my parents of the prognosis. The struggle was particularly intense whenever I saw my father in pain and especially when he would ask me why he was still having pain and feeling bloated after the surgery since the surgery was meant to make him well. Eventually my siblings and I had to tell him that he had cancer as the decision was made by us to begin him on "gentle chemotherapy."

My father withdrew emotionally from us. As the days went by, I saw him withdrawing more and more. He stopped speaking and became more withdrawn and quiet. My father did not live for more than six months. He died in April 1998. It has been nineteen years since my father's death and even today, I wonder if we did the right thing. Did we hasten his death by telling him that he had cancer? Did we make him lose hope? Was it necessary for us to tell him the truth for the chemotherapy treatment? Would it have mattered if he underwent chemotherapy without knowing the truth that he had cancer?

I do not have the answers to my questions. However, I do understand now the struggles that families go through of knowing the truth and trying to protect their loved ones who are diagnosed with a terminal disease. I understand that when they make the decision to withhold the truth, it is one of the most difficult and painful things they must do and it is done with love and not with the malicious intent to deceive.

8.5.1   Introduction

Veracity is a fundamental virtue in nursing. From the time we begin our nursing education to the time we embark on our nursing careers, telling the truth, maintaining integrity, and being ethical form the cornerstone of our professional work of caring. A recent Gallup survey showed that the American public rated nurses as the most ethical and honest healthcare providers among all the other professions [77]. Indeed, nurses have been at the top of the list since 1999.[1] For Singapore nurses, these values, including respecting patients' rights, are enshrined in the *Standards of Practice for Nurses and Midwives* [78] and the *Code of Ethics and Professional Conduct* [79]. Graduating nurses recite the *Nurses Pledge* to publicly affirm they

---

[1] The only year that nurses did not top the list was in 2001 when firefighters were included in response to their work after September 11 attack. Ninety percent of Americans rated firefighters as "high" or "very high" for honesty and integrity. Nurses came in close second at 84% [91].

will uphold the integrity of the profession at all times and make a solemn promise to protect patients' dignity [80]. However, veracity in the form of truth-telling is a double-edged sword—on the one hand, telling the truth is absolutely necessary to build trust and maintain therapeutic relationship with our patients; on the other hand, truth telling may cause ethical dilemmas and/or moral distress among nurses because disclosing the truth is not a simple matter of speaking the truth. Nowhere is this truism more real than in situations when breaking devastating news to patients. Being truthful requires empathy, cultural sensitivity, and a knowledge of individual's values and beliefs. Healthcare providers rooted in Western bioethics consider telling the truth fundamental for a person to exercise her[2] autonomy—that full information must be given to empower a person to make decisions; but from the point of view of individuals who do not share this belief, revealing the truth, especially when confronted with a terminal disease, may be seen as uncaring and uncompassionate [81–87].

In this chapter, we explore the issue of moral distress and truth telling in the Singapore context. Unlike existing literature on moral distress and truth telling, our examination of both concepts is unique because we explore them within the personal realm. We ask the question—would a nurse experience moral distress when confronted with the task of telling the truth to a family member diagnosed with a terminal disease when she is the family caregiver? Rassin [88] believes that the answer lies in understanding the core values of the nurse because values define human behavior and influence choices. He observes that individuals use their values to take a particular stance on specific social and moral issues and people use values to rationalize their behaviors. A nurse, however, operates within three value systems—first, her own unique sociocultural tradition; second, within the nursing value system; and third, within patients' social worldview. For example, both Japanese and American nurses value truth telling to patients; but how each group approach and practice truth telling differs—Japanese nurses prefer to speak the truth using a symbolic and metaphorical language to allow patients to create layered meanings for themselves whereas American nurses prefer truth to be told in a clear and direct manner [89, 90].[3]

8.5.2 Truth-telling and Moral Distress in the Singapore Context

Jameton [92] (p. 297) defines moral distress as "when individuals have clear moral judgments about societal practices, but have difficulty in finding a venue in which to express concerns." Kälvemark et al. expanded Jameton's definition of

---

[2] For the sake of space and to reduce repetition, we will use she as a reference for the person; however, readers should note that this is not an act of discrimination against any gender.

[3] Full disclosure to patients is a recent phenomenon in American medicine. Sisk, Frankel, Kodish, and Isaacson (2016) trace key developments for the past two centuries in American medicine that led to the shift in medical practice—from benevolent paternalism where physicians withheld unpleasant or bad news from their patients due to concerns for patients' well-being to the current practice of transparency and full disclosure. The authors note that even today, physicians struggle to find "the best way to share difficult information without causing undue harm to their patients" (p. 74) [95].

moral distress as "traditional negative stress symptoms that occur due to stress situations that involve ethical dimensions and where the healthcare provider feels that she/he is not able to preserve *all* interests and values at stake" [17] (p. 1082–83; emphasis mine). The expanded definition of moral distress by Kälvemark et al. recognizes that nurses operate in a complex and interconnected environment where there are multiple and competing demands, expectations, and realities. This means that the root of a nurse's moral distress may not only be at the personal level when she is unable to fulfill her moral obligations but her moral distress could be triggered when she perceives that she had failed to meet the needs of other stakeholders. Austin, Lemermeyer, Goldberg, Bergum, and Johnson [93] make similar observations on the ethics of nurse–patient relationships. They contend that because nurses consider their relationship with their patients as privileged, all actions that flow from this privileged relationship are morally defined and as such, when a nurse is unable to fulfill the moral obligations to her patient, the distress becomes apparent. However, Hamric [94] notes that moral distress is a subjective experience because of the multiplicity of values and beliefs held among healthcare providers. While diverse core values, beliefs, and obligations are critical in an inclusive society, these variations could also be a source of tension and conflict because Hamric observes there may not be a consensus of what constitutes correct moral actions since it is difficult to reconcile deeply held values and beliefs. This is the case of Singapore.

Singapore's predominantly Asian, multicultural society[4] provides a fertile ground for this tension and conflict to play out. The government encourages various ethnic groups to practice and maintain their cultural heritage. Values such as filial piety, respect for elders, collective good, and maintaining family harmony are highly esteemed in the Asian context. At the same time, the government urges its citizens to forge a "Singaporean identity"; however, the "Singaporean identity" is a fluid concept and is work in progress.[5] Overarching the multicultural framework are two other key systems. First, Singapore's laws are based on English law, a legacy of the country's colonial past. Chan [96] observes that over the years, Singapore has adopted and adapted the English common law into its own set of laws, and that these laws have shaped the country's various political, social, and economic institutions.

---

[4] The multicultural framework referred here is the official Chinese, Malay, Indian, and Others (CMIO) categorization that Singapore uses in all its official communications and government publications. Within the nursing community, it will be seen that the category of Others has expanded. For example, we now have significant proportion of nurses from the Philippines, China, India, Malaysia, and Myanmar (SNB, 2015). Singapore Nursing Board (2015). *Annual report.* Singapore: SNB. Retrieved from http://www.healthprofessionals.gov.sg/content/dam/hprof/snb/docs/publications/SNB%20Annual%20Report%202015_%2030%20Aug%202016.pdf.

[5] Singapore became independent in 1965 and except for the indigenous population, Singapore is a land of immigrants whose ancestors came from China, India, Middle East, and Indonesia (Chan, 2013). Immigration continues even today as the country's total fertility rate (TFR) is low—it was 1.2 in 2016 (Singapore Statistics, 2017). Department of Statistics (2017). *Births and deaths.* Singapore: Government of Singapore. Retrieved from http://www.singstat.gov.sg/statistics/latest-data#.

Second, and related to its colonial legacy, is a healthcare system that is deeply entrenched in the biomedical ethos. This has led to the creation of a distinctive social culture where both Asian and Western values thrive in Singapore, and where individuals learn to weave in and out seamlessly and live with contradictions. Singapore's unique culture also means that there are many permutations of truth telling. It is within these layered frameworks that nurses navigate to provide culturally appropriate care to patients. Pergert and Lützén [85] and others [97, 98] remind us that truth telling is very much a cultural artifact and as such, there is no one correct way of stating the truth. They [85] suggest that instead of solely focusing on autonomy as the main guiding principle in truth telling, the nurse should consider relational ethics to explore the concept of truth telling with her patient. Relational ethics allows for "ethical practice [to be] situated in relationships" implying that the artificial demarcation of independency does not reflect the reality since all of us are born into relationships [85] (p. 25)

Truth telling is a complex undertaking because a nurse in Singapore negotiates between her personal and professional values with patient's needs and expectations and the family's wishes. Tension exists because the nurse must balance various obligations without compromising care or violating her professional ethics. It is common to see patients deferring to their families when it comes to making important decisions regarding their health and treatment. Families play an active role to control the level and type of information given to their loved ones to protect them from losing hope [87]. Unlike in a Western[6] society, where individual autonomy is central to exercising one's rights and choices, in the Singapore context, the family exerts a powerful and often direct influence on patient's choices. Tan and Chin found that doctors often colluded with family members to decide whether to inform the patient of her diagnosis and what information to divulge. The authors [87] learned that this caused moral distress among the physicians and the latter worked hard to convince family members that the patient should know her prognosis. At the same time, physicians respected the families' decision not to reveal the truth because in most instances, they understood that families were trying to protect their loved ones from losing hope:

> "Family members who try to prevent disclosure by doctors to patients, and who try to make decisions for patients, appear to be doing so out of good intentions. All the doctors who described such actions said that relatives were well intentioned in their actions. The relatives were generally trying to protect the patients from the burden of knowledge, the burden of responsibility of making decisions, and in particular the prospect of 'losing hope' if given bad news" [87] (p. 14).

In Singapore, the intersection of biomedical culture and the patient's sociocultural worldview become points of contention and conflict when the individual enters the healthcare system and assumes the role of a patient. The nurse, governed by the

---

[6]We note that the term used here suggest that the West is a homogenous and monolithic. We acknowledge that this is not the case and that even in a "Western" society, there are multiple realities and complexities. The homogeneity we are referring to is the dominant and influential Judeo Christian culture that influences the various institutions and thinking.

professional code of ethics and the rule of law, must negotiate through these two worldviews in addition to balancing her own values and beliefs and meeting the needs of the patient. Although the nurse operates strictly on the first two world-views—advocating and promoting patient's interests and fulfilling her professional obligations, she is not completely divorced from her own values. Indeed, it would be unrealistic to expect the nurse to be completely distanced from her own personal value system even when she is fulfilling her professional obligations.

In most instances, there is a peaceful co-existence between the biomedical and the patient's unique sociocultural systems. The clash occurs when one of the two systems attempts to exert its influence on the other to determine the greater good for the patient. Although Singapore's unique social fabric allows two parallel value systems to exist and continue within the healthcare structure, it is often the biomedical healing system that dominates and wields its authoritative knowledge since it is perceived to be objective and scientific. Patient's cultural and social values are accommodated only insofar as these values do not question the biomedical authority or interfere with its processes. And yet there is a good evidence that indicates that providing culturally congruent care that aligns with patient's values, beliefs, and practices has tremendous benefits for patients' well-being [98–103]. Moral distress arises when there is a mismatch between congruency of care and the nurse professional ethics and values. Epstein and Hamric [104] observe that if the root cause of the nurse moral distress is not addressed satisfactorily, the nurse may experience the lingering effect of the moral distress known as moral residue. Over time, the cumulative effects of moral residue and moral distress could lead to the crescendo effect, a more serious outcome of moral distress [104].

8.5.3   Moral Distress in the Personal Space: The Nurse as the Family Caregiver

The discussion on moral distress so far has centered on the nurse's professional space, focusing on the dissonance between the nurse's moral obligations and patient's values and beliefs. In this section, we want to shift the discussion on moral distress within the personal context of the nurse. At the beginning, we had asked the question *whether a nurse would experience moral distress when confronted with the task of telling the truth to a family member diagnosed with a terminal disease when she is the family caregiver.* This is the situation that one of us (Subadhra)[7] found herself when faced with the task of telling the truth to her father diagnosed with pancreatic cancer. The answer to this question is yes because the personal context brings its own set of challenges. Unlike the professional space, where clear boundaries and practices exist to assist nurses to provide therapeutic care safely, the personal context is fraught with potential conflicts because the line between professional obligations and personal values and expectations is often blurred.

---

[7] I will speak in the first person.

My moral distress stemmed from my own inner battle—to which parent do I owe a duty of care to tell the truth about the diagnosis and the prognosis? Do I owe a duty of care to my father to inform him that his cancer had metastasized and the surgery was non-therapeutic? Or, do I owe a duty to my mother to prepare her of the impending change in her social status—from being a wife to a widow? Moreover, in all this, which role should attain primacy—the daughter or the nurse? From the perspective of a daughter, I felt that I owed a duty of care to both of my parents. I wanted to protect both from the painful news of eventual death and separation. I wanted to buffer them as much as I could to delay the burden of grieving. In particular, I wanted to protect my mother of her eventual role change from a wife to a widow because in the South Asian context, being a widow has profound consequences on the woman in the community. Apart from the physical separation, it is the various symbolic separations that reinforce her declined standing in the community.[8] Knowing the impact that the death of my father would have on my mother, I wanted to delay the grief from the knowledge; however, I also knew that eventually I could not shield her from the pain of her deep loss. As a daughter, I owed a duty of care to my father to be honest with him and to be his advocate. But I could not do either as I feared that he would lose hope once he knew he had cancer and as noted earlier, I wished to protect him from the devastating news. And yet, deep within me, I knew that knowing the truth could enable both of my parents to "deal" with what was to come.

The moral distress I experienced as a daughter occurred because I let both of my parents down by not facilitating the truth with compassion. I began to question whether what I was doing was truly for my parents or was it due to my own fear of the imminent loss. I must admit that my actions and feelings as a daughter and that of a nurse began to merge and became indistinguishable. Often both perspectives (daughter and nurse) occurred simultaneously and during these moments, I would experience the intense moral distress because both of my value systems collided. The daughter part of me considered that withholding the truth and making decisions on behalf of my parents was compassionate care and I was fulfilling my role as a filial daughter; however, my training as a nurse and the values that I had embraced considered my actions as unethical and definitely paternalistic. Begley [82] and Pergert and Lützén [85] make the point that pushing for an absolutist approach to truth telling or for patient's autonomy is unhelpful; instead, the authors recommend that autonomy and truth telling be contextualized within the larger social and cultural fabric of the patient and the family. They contend that there may be instances

---

[8] Times have changed and some of these practices are no longer practiced by the women (descendants of the earlier immigrants and newcomers of Uttar Pradesh [U.P]) in Singapore. However, for women of my mother's generation, many of whom immigrated from India, these social and cultural mores were strictly adhered to within our U.P. community in Singapore. Widowhood was not merely a physical separation of a woman from her partner but each symbolic action reinforced her declined social standing. These include observing certain food prohibitions to removing the vermillion *tikka* on her forehead to removing her colored bangles to wearing a white sari to unable to attend certain auspicious functions. Repeatedly, my mother was reminded of what she had been when my father was alive to what she is now when my father died of cancer.

where implementing full autonomy may not be possible and withholding the telling of the truth would be an act of compassion [82, 85, 97, 98].

When I viewed the situation from the perspective of a nurse, it was clear to me what my obligations were to my father. I strongly believed that he had the right to know about his health and decide which treatment options to choose instead of his children deciding for him. It was clear to me that my duty was to advocate for his autonomous decision-making and that deciding for him was an act of paternalism on my part. Therefore, withholding the truth from my parents, and from my father, was unethical even though I understood the good intention. However, this knowledge of good did not assuage the anger and disappointment I felt of my own futility and inability to speak up for my father or persuade the telling of the truth in an empathetic way. Hamric sums this distress well when she says, "moral distress can cut to the heart of one's view of oneself as a moral professional and moral person" [94] (p. 457).

How did I deal with my moral distress? There is no clear answer to this question because the options available to me—withholding or telling the truth—were never easy choices to make. Rassin [88] is correct when he states that individuals' core values and beliefs may guide how they resolve their moral distress and dilemmas. In the end, I relied on my cultural values[9] of being filial, promoting collective good, and respecting the culture of my parents. I used my sociocultural belief to rationalize the withholding of truth. For my siblings and me, compassion and kindness overrode the issue of honesty and autonomy. In fact, the roles between my parents and us became reversed—we became the "parents" and my parents, the "children" who needed care and empathy. The two guiding principles—doing my duty (*Dharma*) and the consequences of my actions (*Karma*) were fundamental in how I rationalized my decision to withhold the telling of the truth. I felt that withholding the truth was justified because it was for the collective good of my parents instead of pushing for my father's autonomy.

Eventually we had to inform our parents of my father's diagnosis. My siblings and I discussed that we should at least give our father the opportunity to decide whether he wished to have chemotherapy. However, we could not speak the whole truth, as we did not wish our father to despair on hearing that he had a cancer which was terminal. This was the second time when I experienced moral distress—the inner turmoil of knowing that I ought to give my father the complete truth but unable to do so as I feared that the news would devastate my parents further. I had to do self-rationalization for the second time—that a half-truth is better than withholding the truth. I realized that my training as a nurse was a double-edged sword. My nursing knowledge and skills gave me the confidence to nurse my father at home and to manage his pain and ensure a 'good death.' But, nursing was also a source of pain and distress for me as I was aware of what was the right thing to do but unable to do so to the full extent.

---

[9] I was raised in the Hindu tradition.

8.5.4 Final Thoughts…

My father's story is a story that repeats every day in Singapore. Families shelter their loved ones from the knowledge of hearing devastating news so that the person will not lose hope but finds the inner strength to deal with the illness. The assumption behind this belief is news of any illness (especially long-term) creates a sense of disequilibrium in the individual's worldview; and compounding this disequilibrium with bad news would further undermine the person's coping capacity. Supporters of autonomy, however, point out that this line of thinking does not give credit to the individual's right to make a decision nor acknowledges the person's inner resilience to soldier on even when receiving bad news such as a terminal disease. This criticism may well be deserved but for families who are dealing with the knowledge of eventual loss, autonomy and rights are abstract terms. Compassion and love are the values that motivate families to bear the emotional burden of withholding the truth, recognizing that an individual is part of the larger human network and interconnected to others.

As nurses, we were especially interested to explore whether a nurse experienced moral distress during truth telling when she was the family caregiver. We learned that the nurse dual role—a family member and a nurse—creates a fertile ground for moral distress and moral ambiguities. The distress arises when the nurse knows what she ought to do—facilitate and advocate for truth telling with, and for, the family member but is unable to do so because the sociocultural and spiritual values of the family and hers may not allow for complete truth telling to the patient. The nurse, who is both a family member and the family caregiver, must negotiate between two value systems—the professional code of ethics which states clearly her ethical obligations to her patients, and the personal—the sociocultural system which anchors on an existential philosophical foundation that allows people to create personal meanings. The moral ambiguity emerges when these two value systems compete and both have equal moral authority. We believe that Kälvemark et. al.'s [17] definition of moral distress captures the true essence of nurse moral distress because the nurse tries to fulfill *all* her ethical obligations and finds that she is unable to do so to the full extent.

Suggested Readings

Brenan M. Nurses keep healthy lead as most honest, ethical profession. Gallup. 2017. Retrieved from http://www.news.gallup.com/poll/224639/nurses-keep-healthy-lead-honest-ethicalprofession.aspx?.

Singapore Nursing Board. Standards of practice for nurses and midwives. Singapore: SNB. 2011. Retrieved from http://www.healthprofessionals.gov.sg/content/dam/hprof/snb/docs/publications/SNB_Standards_of_Practices.pdf.

Singapore Nursing Board. Code of ethics and professional conduct. Singapore: SNB. 1999. Retrieved from http://www.healthprofessionals.gov.sg/content/dam/hprof/snb/docs/publications/Code%20of%20Ethics%20and%20Professional%20Conduct%20(15%20Mar%20 1999).pdf.

Singapore Nursing Board. Nurses' pledge. Singapore: SNB. n.d. Retrieved from http://www. healthprofessionals.gov.sg/content/dam/hprof/snb/docs/publications/Nurses%20Pledge.pdf.

Begley A, Blackwood B. Truth-telling versus hope: a dilemma in practice. Int J Nurs Pract. 2000;6(1):26–31.

Begley AM. Truth-telling, honesty and compassion: a virtue-based exploration of a dilemma in practice. Int J Nurs Pract. 2008;14(5):336–41.

Guven T. Truth-telling in cancer: examining the cultural incompatibility argument in Turkey. Nurs Ethics. 2010;17(2):159–66.

Lee A, Wu HY. Diagnosis disclosure in cancer patients – when the family says "no!". Singap Med J. 2002;43(10):533–8.

Pergert P, Lüzén K. Balancing truth-telling in the preservation of hope: a relational ethics approach. Nurs Ethics. 2012;19(1):21–9.

Slowther A. Truth-telling in health care. Clin Ethics. 2009;4(4):173–5.

Tan JOA, Chin JJL. What doctors say about care of the dying. Singapore: The Lien Foundation. 2011. Retrieved from http://www.lienfoundation.org/sites/default/files/What_Doctors_Say_About_Care_of_the_Dying_0.pdf.

Rassin M. Nurses' professional and personal values. Nurs Ethics. 2008;15(5):614–30.

Izumi S. Bridging Western ethics and Japanese local ethics by listening to nurses' concerns. Nurs Ethics. 2006;13(3):275–83.

Wros PL, Doutrich D, Izumi S. Ethical concerns: comparison of values from two cultures. Nurs Health Sci. 2004;6(2):131–40.

Moore, D. W. (2001, December 5). Firefighters top Gallup's "honesty and ethics" list. CNN/USA Today/Gallup Poll. Retrieved from http://www.gallup.com/poll/5095/Firefighters-Top-Gallups-Honesty-Ethics-List.aspx.

Jameton A. A reflection on moral distress in nursing together with a current application of the concept. J Bioethical Inq. 2013;10(3):297–308.

Kalvemark S, Hoglund AT, Hansson MG, Westerholm P, Arnetz B. Living with conflicts-ethical dilemmas and moral distress in the health care system. Soc Sci Med. 2004;58(6):1075–84.

Austin W, Lemermeyer G, Goldberg L, Bergum V, Johnson MS. Moral distress in healthcare practice: the situation of nurses. HEC Forum. 2005;17(1):33–48.

Hamric AB. A case study of moral distress. J Hospice Palliat Care. 2014;16(8):457–63.

Sisk et al. The truth about truth-telling in American medicine: A brief history. The Permanente Journal. 2016;20(3):74–7.

Chan SK. Multiculturalism in Singapore: the way to a harmonious society. Singapore Acad Law J. 2013;23:84–109. Retrieved from http://journalsonline.academypublishing.org.sg/Journals/Singapore-Academy-of-Law-Journal/e-Archive/ctl/eFirstSALPDFJournalView/mid/495/ArticleId/500/Citation/JournalsOnlinePDF

Leever MG. Cultural competence: reflections on patient autonomy and patient good. Nurs Ethics. 2011;18(4):560–70.

Rising ML. Truth telling as an element of culturally competent care at end of life. J Transcult Nurs. 2017;28(1):48–55.

Betancourt JR, Green AR, Carrillo JE, Park ER. Cultural competence and health care disparities: Key perspectives and trends. Health Affairs. 2005;24(2):499–505.

Campinha-Bacote J. The process of cultural competence in the delivery of healthcare services: A model of care. Journal of Transcultural Nursing. 2002;13(3):181–4.

Milton CL. Ethics and defining cultural competence: An alternative view. Nursing Science Quarterly, 2016;29(1):21–3.

Saha S, Beach ML, Cooper LA. Patient centeredness, cultural competence and healthcare quality. Journal of National Medical Association. 2008;100(11):1275–85.

Stewart S. Cultural competence in healthcare. Diversity Health Institute. Position Paper. 2006. Retrieved from http://citeseerx.ist.psu.edu/viewdoc/download?doi=10.1.1.110.8602&rep=rep1&type=pdf.

Epstein EG, Hamric AB. Moral distress, moral residue and the crescendo effect. J Clin Ethics. 2009;20(4):330–42.

References

1. Lamiani G, Borghi L, Argentero P. When healthcare professionals cannot do the right thing: a systematic review of moral distress and its correlates. J Health Psychol. 2017;22(1):51–67.
2. Atabay G, Cangarli BG, Penbek S. Impact of ethical climate on moral distress revisited: multidimensional view. Nurs Ethics. 2015;22(1):103–16.
3. Hurst SA, Perrier A, Pegoraro R, Reiter-Theil S, Forde R, Slowther AM, et al. Ethical difficulties in clinical practice: experiences of European doctors. J Med Ethics. 2007;33(1):51–7.
4. Papathanassoglou ED, Karanikola MN, Kalafati M, Giannakopoulou M, Lemonidou C, Albarran JW. Professional autonomy, collaboration with physicians, and moral distress among European intensive care nurses. Am J Crit Care. 2012;21(2):e41–52.
5. Corley MC, Elswick RK, Gorman M, Clor T. Development and evaluation of a moral distress scale. J Adv Nurs. 2001;33(2):250–6.
6. Silen M, Svantesson M, Kjellstrom S, Sidenvall B, Christensson L. Moral distress and ethical climate in a Swedish nursing context: perceptions and instrument usability. J Clin Nurs. 2011;20(23-24):3483–93.
7. Goethals S, Gastmans C, de Casterle BD. Nurses' ethical reasoning and behaviour: a literature review. Int J Nurs Stud. 2010;47(5):635–50.
8. Shorideh FA, Ashktorab T, Yaghmaei F. Iranian intensive care unit nurses' moral distress: a content analysis. Nurs Ethics. 2012;19(4):464–78.
9. de Veer AJ, Francke AL, Struijs A, Willems DL. Determinants of moral distress in daily nursing practice: a cross sectional correlational questionnaire survey. Int J Nurs Stud. 2013;50(1):100–8.
10. Piers RD, Azoulay E, Ricou B, DeKeyser Ganz F, Max A, Michalsen A, et al. Inappropriate care in European ICUs: confronting views from nurses and junior and senior physicians. Chest. 2014;146(2):267–75.
11. de Boer J, van Rosmalen J, Bakker A, van Dijk M. Appropriateness of care and moral distress among neonatal intensive care unit staff: repeated measures. Nurs Crit Care. 2015;21:19–27.
12. Oerlemans AJ, van Sluisveld N, van Leeuwen ES, Wollersheim H, Dekkers WJ, Zegers M. Ethical problems in intensive care unit admission and discharge decisions: a qualitative study among physicians and nurses in the Netherlands. BMC Med Ethics. 2015;16:9.
13. Sannino P, Gianni ML, Re LG, Lusignani M. Moral distress in the neonatal intensive care unit: an Italian study. J Perinatol. 2015;35(3):214–7.
14. Lusignani M, Gianni ML, Re LG, Buffon ML. Moral distress among nurses in medical, surgical and intensive-care units. J Nurs Manag. 2016;25(6):477–85.
15. Cuttini M, Nadai M, Kaminski M, Hansen G, de Leeuw R, Lenoir S, et al. End-of-life decisions in neonatal intensive care: physicians' self-reported practices in seven European countries. EURONIC Study Group. Lancet. 2000;355(9221):2112–8.
16. Sprung CL, Cohen SL, Sjokvist P, Baras M, Bulow HH, Hovilehto S, et al. End-of-life practices in European intensive care units: the Ethicus Study. JAMA. 2003;290(6):790–7.
17. Kalvemark S, Hoglund AT, Hansson MG, Westerholm P, Arnetz B. Living with conflicts-ethical dilemmas and moral distress in the health care system. Soc Sci Med. 2004;58(6):1075–84.
18. Karanikola MN, Albarran JW, Drigo E, Giannakopoulou M, Kalafati M, Mpouzika M, et al. Moral distress, autonomy and nurse-physician collaboration among intensive care unit nurses in Italy. J Nurs Manag. 2014;22(4):472–84.
19. Benbenishty J, Ganz FD, Lippert A, Bulow HH, Wennberg E, Henderson B, et al. Nurse involvement in end-of-life decision making: the ETHICUS Study. Intensive Care Med. 2006;32(1):129–32.
20. Jensen HI, Ammentorp J, Erlandsen M, Ording H. Withholding or withdrawing therapy in intensive care units: an analysis of collaboration among healthcare professionals. Intensive Care Med. 2011;37(10):1696–705.

21. Oh Y, Gastmans C. Moral distress experienced by nurses: a quantitative literature review. Nurs Ethics. 2015;22(1):15–31.
22. Forde R, Aasland OG. Moral distress among Norwegian doctors. J Med Ethics. 2008;34(7):521–5.
23. Lievrouw A, Vanheule S, Deveugele M, Vos M, Pattyn P, Belle V, et al. Coping with moral distress in oncology practice: nurse and physician strategies. Oncol Nurs Forum. 2016;43(4):505–12.
24. Anstey MH, Adams JL, McGlynn EA. Perceptions of the appropriateness of care in California adult intensive care units. Crit Care. 2015;19:51.
25. Piers RD, Azoulay E, Ricou B, Dekeyser Ganz F, Decruyenaere J, Max A, et al. Perceptions of appropriateness of care among European and Israeli intensive care unit nurses and physicians. JAMA. 2011;306(24):2694–703.
26. Musto LC, Rodney PA, Vanderheide R. Toward interventions to address moral distress: navigating structure and agency. Nurs Ethics. 2015;22(1):91–102.
27. Reader TW, Flin R, Mearns K, Cuthbertson BH. Interdisciplinary communication in the intensive care unit. Br J Anaesth. 2007;98(3):347–52.
28. Van den Bulcke B, Vyt A, Vanheule S, Hoste E, Decruyenaere J, Benoit D. The perceived quality of interprofessional teamwork in an intensive care unit: a single centre intervention study. J Interprof Care. 2016;30(3):301–8.
29. Quenot JP, Rigaud JP, Prin S, Barbar S, Pavon A, Hamet M, et al. Suffering among carers working in critical care can be reduced by an intensive communication strategy on end-of-life practices. Intensive Care Med. 2012;38(1):55–61.
30. Hartog CS, Schwarzkopf D, Riedemann NC, Pfeifer R, Guenther A, Egerland K, et al. End-of-life care in the intensive care unit: a patient-based questionnaire of intensive care unit staff perception and relatives' psychological response. Palliat Med. 2015;29(4):336–45.
31. Bruce CR, Miller SM, Zimmerman JL. A qualitative study exploring moral distress in the ICU team: the importance of unit functionality and intrateam dynamics. Crit Care Med. 2015;43(4):823–31.
32. Piquette D, Reeves S, LeBlanc VR. Stressful intensive care unit medical crises: how individual responses impact on team performance. Crit Care Med. 2009;37(4):1251–5.
33. Kalvemark Sporrong S, Arnetz B, Hansson MG, Westerholm P, Hoglund AT. Developing ethical competence in health care organizations. Nurs Ethics. 2007;14(6):825–37.
34. Hough CL, Hudson LD, Salud A, Lahey T, Curtis JR. Death rounds: end-of-life discussions among medical residents in the intensive care unit. J Crit Care. 2005;20(1):20–5.
35. Murray SA, Kendall M, Boyd K, Sheikh A. Illness trajectories and palliative care. BMJ. 2005;330(7498):1007–11.
36. Jameton A. Nursing practice: the ethical issues. New Jersey: Prentice Hall; 1984.
37. Truog RD, Brown SD, Browning D, Hundert EM, Rider EA, Bell SK, et al. Microethics: the ethics of everyday clinical practice. Hastings Cent Rep. 2015;45(1):11–7.
38. Nursing & Midwifery Council N. The Code: professional standards of practice and behaviour for nurses and midwives: Nursing & Midwifery Council N. 2015.
39. Nursing & Midwifery Council N. Standards for competence for registered nurses. London: Nursing & Midiwfery Council, NMC. 2014. p. 1–21.
40. Nursing & Midwifery Council N. Standards for pre-registration nursing education. London: Nursing & Midwifery Council, NMC. 2010. p. 1–152.
41. Hamric AB, Arras JD, Mohrmann ME. Must we be courageous? Hast Cent Rep. 2015;45(3):33–40.
42. Tessman L. Burdened virtues: virtue ethics for libertory struggles. New York: Oxford Scholarship Online; 2005.
43. Hamric A, Wocial LD. Institutional ethics resources: creating moral spaces. Hast Cent Rep. 2016;46(Suppl 1):S22–7.
44. Walker MU. Keeping moral space open: new images of ethics consulting. Hast Cent Rep. 1993;23(2):33–40.

45. Traudt T, Liaschenko J, Peden-McApline C. Moral agency, moral imagination, and moral community: antidotes to moral distress. J Clin Ethics. 2016;27(3):201–13.
46. Morley G. Efficacy of the nurse ethicist in reducing moral distress: what can the NHS learn from the USA? Part 2. Br J Nurs. 2016;25(3):156–61.
47. Warren VL. Feminist directions in medical ethics. Hypatia. 1989;4(2):73–87.
48. Fitzpatrick P, Scully JL. Theory in feminist bioethics. In: Scully JL, Baldwin-Ragaven LE, Fitzpatrick P, editors. Feminist bioethics at the center, on the margins. Baltimore: Johns Hopkins University Press; 2010. p. 61–9.
49. Kittay EF, Jennings B, Wasunna AA. Dependency, difference and the global ethic of longterm care. J Polit Philos. 2005;13(4):443–69.
50. Whitehead PB, Herbertson RK, Hamric AB, Epstein EG, Fisher JM. Moral distress among healthcare professionals: report of an institution-wide survey. J Nurs Scholarsh. 2014;47(2):117–25. https://doi.org/10.1111/jnu.12115.
51. Lindemann H. Speaking truth to power. Hast Cent Rep. 2010;40(1):44–5. https://doi.org/10.1353/hcr.0.0215.
52. Walker MU. Moral understandings: a feminist study in Ethics. 2nd ed. New York: Oxford University Press; 2007.
53. Gotlib A. Feminist ethics and narrative ethics. Internet encyclopedia of philosophy. 2014. http://www.iep.utm.edu/fem-e-n/. Accessed 11 Nov 2016.
54. Walker MU. Introduction: Groningen naturalism in bioethics. In: Lindemann H, Verkerk M, Walker MU, editors. Naturalized bioethics: toward responsible knowing and practice. New York: Cambridge University Press; 2009. p. 1–20.
55. Tanzanian and Midwifery iCouncil. Code of Ethics and Professional Conduct for Nurses and Midwives in Tanzania Revised. Tanzania; 2015.
56. Elpern B, Covert B, Kleinpell R. Moral distress of staff nurses in a medical intensive care unit. Am J Crit Care. 2005;14(6):523–30.
57. Sirili N, Kiwara A, Nyongole O, Frumence G, Semakafu A, Hurtig AK. Addressing the human resource for health crisis in Tanzania: the lost in transition syndrome. Tanzan J Health Res. 2014;16(2):1–9.
58. Sikika, Medical Association Of Tanzania. Where are the Doctors? - Tracking Study of Medical Doctors. 2013. http://sikika.or.tz/wp-content/uploads/2014/03/Practice-Status-of-Medical-Graduates-MD-Tracking-edited.pdf.
59. Medical Association of Tanzania (MAT). Proceedings of the 43rd Annual General Meeting and 45th Anniversary. 2010.
60. Kwesigabo G, Mwangu MA, Kakoko DC, Warriner I, Mkony CA, Killewo J, et al. Tanzania's health system and workforce crisis. J Public Health Policy. 2012;33(Suppl 1(S1)):S35–44. http://www.ncbi.nlm.nih.gov/pubmed/23254848
61. The United Republic of Tanzania Ministry of Health and Social Welfare. Human Resource for Health Strategic Plan 2008–2013. Minist Heal Soc Welf. 2008.
62. Hellsten SK. Bioethics in Tanzania: legal and ethical concerns in medical care and research in relation to the HIV/AIDS epidemic. Camb Q Healthc Ethics. 2005;2005:256–67.
63. Economic and Social Research Foundation. Development report, 2014; Economic Transformation for Human Development. 2014. p. 1–128.
64. Hamric AB, Davis WS, Childress MD. Moral distress in health care professionals. Pharos Alpha Omega Alpha Honor Med Soc. 2006;69(1):16–23. http://www.ncbi.nlm.nih.gov/pubmed/16544460
65. Kuwawenaruwa A, Borghi J. Health insurance cover is increasing among the Tanzanian population but wealthier groups are more likely to benefit. Ifakara Health Institute. 2012. p. 1–4.
66. Vallely A, Lees S, Shagi, C., Kasindi S, Soteli S, Kavit N, Vallely L, McCormack S, Pool R, Hayes RJ and the Microbicides Development Programme (MDP). How informed is consent in vulnerable populations? Experience using a continuous consent process during the MDP301 vaginal microbicide trial in Mwanza, Tanzania. BMC Med Ethics 2010; 11:10

67. Manzi F, Schellenberg J, Hutton G, Wyss K, Mbuya C, Shirima K, et al. Human resources for health care delivery in Tanzania: a multifaceted problem. Hum Resour Health. 2012;10(1):3. http://www.human-resources-health.com/content/10/1/3

68. National Bureau of Statistics (NBS). The United Republic of Tanzania 2015 Tanzania in figures. 2016. http://www.nbs.go.tz/nbs/takwimu/references/Tanzania_in_Figures_2015.pdf.

69. Tibandebage BP, Kida T, Mackintosh M, Ikingura J. Understandings of ethics in maternal health care: an exploration of evidence from four districts in Tanzania. 2013. http://www.repoa.or.tz/documents/REPOA_WORKING_PAPER_13.2.pdf.

70. Kwesigabo G, Mwangu MA, Kakoko DC, Killewo J. Health challenges in Tanzania: context for educating health professionals. J Public Health Policy. 2012;33(Suppl 1):S23–34.

71. Kurowski C, Wyss K, Abdulla S, Yémadji D, Mills A. Human resources for health: requirements and availability in the context of scaling-up priority interventions in low-income countries Case studies from Tanzania and Chad. Heal Econ Financ Program. 2004. p. 96. https://assets.publishing.service.gov.uk/media/57a08c12e5274a31e0000f9a/WP01_04.pdf.

72. Munga MA, Maestad O. Measuring inequalities in the distribution of health workers: the case of Tanzania. Hum Resour Health. 2009;7:4.

73. Education MOF, Training V, Technology C, Syllabus A, Certificate FOR, In C, et al. the United Republic of Tanzania. 2009.

74. Ulrich CM, Muecke, M., Mann Wall, B., Hoke, L., Joseph, R., Shayo, J.E., Morris, B.M., Sabone, M., Cainelli, F., Mazonde, P., Maitshoko, M. Inter-professional practice and education: a collaborative initiative with Tanzania and Botswana: Penn in Africa. 2013.

75. Waddell R, Aboud M. Dartmouth/Muhas research ethics training and program development for Tanzania, [R25TW007693]. Tanzania: National Institutes of Health, Fogarty; 2011-2016.

76. Ringer S, Aboud M. Dartmouth/Muhas Research Ethics Training and Program Development for Tanzania. [R25TW007693]; Tanzania: National Institutes of Health, Fogarty; 2017-2022. National Institutes of Health, Fogarty; 2017-2022.

77. Brenan M. Nurses keep healthy lead as most honest, ethical profession. Gallup. 2017. Retrieved from http://news.gallup.com/poll/224639/nurses-keep-healthy-leadhonest-ethical-profession.aspx?.

78. Singapore Nursing Board. Standards of practice for nurses and midwives. Singapore: SNB. 2011. Retrieved from http://www.healthprofessionals.gov.sg/content/dam/hprof/snb/docs/publications/SNB_Standards_of_Practices.pdf.

79. Singapore Nursing Board. Code of ethics and professional conduct. Singapore: SNB. 1999. Retrieved from http://www.healthprofessionals.gov.sg/content/dam/hprof/snb/docs/publications/Code%20of%20Ethics%20and%20Professional%20Conduct%20(15%20Mar%201999).pdf.

80. Singapore Nursing Board. Nurses' pledge. Singapore: SNB. n.d. Retrieved from http://www.healthprofessionals.gov.sg/content/dam/hprof/snb/docs/publications/Nurses%20Pledge.pdf.

81. Begley A, Blackwood B. Truth-telling versus hope: a dilemma in practice. Int J Nurs Pract. 2000;6(1):26–31.

82. Begley AM. Truth-telling, honesty and compassion: a virtue-based exploration of a dilemma in practice. Int J Nurs Pract. 2008;14(5):336–41.

83. Guven T. Truth-telling in cancer: examining the cultural incompatibility argument in Turkey. Nurs Ethics. 2010;17(2):159–66.

84. Lee A, Wu HY. Diagnosis disclosure in cancer patients – when the family says "no!". Singap Med J. 2002;43(10):533–8.

85. Pergert P, Lüzén K. Balancing truth-telling in the preservation of hope: a relational ethics approach. Nurs Ethics. 2012;19(1):21–9.

86. Slowther A. Truth-telling in health care. Clin Ethics. 2009;4(4):173–5.

87. Tan JOA, Chin JJL. What doctors say about care of the dying. Singapore: The Lien Foundation. 2011. Retrieved from http://www.lienfoundation.org/sites/default/files/What_Doctors_Say_About_Care_of_the_Dying_0.pdf.

88. Rassin M. Nurses' professional and personal values. Nurs Ethics. 2008;15(5):614–30.

89. Izumi S. Bridging Western ethics and Japanese local ethics by listening to nurses' concerns. Nurs Ethics. 2006;13(3):275–83.
90. Wros PL, Doutrich D, Izumi S. Ethical concerns: comparison of values from two cultures. Nurs Health Sci. 2004;6(2):131–40.
91. Moore, D. W. (2001, December 5). Firefighters top Gallup's "honesty and ethics" list. CNN/ USA Today/Gallup Poll. Retrieved from http://www.gallup.com/poll/5095/Firefighters-Top-Gallups-Honesty-Ethics-List.aspx.
92. Jameton A. A reflection on moral distress in nursing together with a current application of the concept. J Bioethical Inq. 2013;10(3):297–308.
93. Austin W, Lemermeyer G, Goldberg L, Bergum V, Johnson MS. Moral distress in healthcare practice: the situation of nurses. HEC Forum. 2005;17(1):33–48.
94. Hamric AB. A case study of moral distress. J Hospice Palliat Care. 2014;16(8):457–63.
95. Sisk et al. (2016). The truth about truth-telling in American medicine: A brief history. *The Permanente Journal, 20*(3),74–77.
96. Chan SK. Multiculturalism in Singapore: the way to a harmonious society. Singapore Acad Law J. 2013;23:84–109. Retrieved from http://journalsonline.academypublishing.org.sg/ Journals/Singapore-Academy-of-Law-Journal/e-Archive/ctl/eFirstSALPDFJournalView/ mid/495/ArticleId/500/Citation/JournalsOnlinePDF
97. Leever MG. Cultural competence: reflections on patient autonomy and patient good. Nurs Ethics. 2011;18(4):560–70.
98. Rising ML. Truth telling as an element of culturally competent care at end of life. J Transcult Nurs. 2017;28(1):48–55.
99. Betancourt JR, Green AR, Carrillo JE, Park ER. Cultural competence and health care disparities: Key perspectives and trends. Health Affairs. 2005;24(2):499–505.
100. Campinha-Bacote J. The process of cultural competence in the delivery of healthcare services: A model of care. Journal of Transcultural Nursing. 2002;13(3):181–4.
101. Milton CL. Ethics and defining cultural competence: An alternative view. Nursing Science Quarterly, 2016;29(1):21–3.
102. Saha S, Beach ML, Cooper LA. Patient centeredness, cultural competence and healthcare quality. Journal of National Medical Association. 2008;100(11):1275–85.
103. Stewart S. Cultural competence in healthcare. Diversity Health Institute. Position Paper. 2006. Retrieved from http://citeseerx.ist.psu.edu/viewdoc/download?doi=10.1.1.110.8602& rep=rep1&type=pdf.
104. Epstein EG, Hamric AB. Moral distress, moral residue, and the crescendo effect. J Clin Ethics. 2009;20(4):330–42.

Christine Grady, Nancy Berlinger, Arthur Caplan,
Sheila Davis, Ann B. Hamric, Shaké Ketefian, Robert Truog,
and Connie M. Ulrich

C. Grady (✉)
Department of Bioethics, National Institutes of Health, Bethesda, MD, USA
e-mail: CGrady@cc.nih.gov

N. Berlinger
The Hastings Center, Garrison, NY, USA

A. Caplan
Division of Medical Ethics, NYU School of Medicine, New York, NY, USA

S. Davis
Partners In Health, Boston, MA, USA

A.B. Hamric
School of Nursing, Virginia Commonwealth University, Richmond, VA, USA

S. Ketefian
School of Nursing, University of Michigan, Ann Arbor, MI, USA

R. Truog
Medical Ethics, Anaesthesia, and Pediatrics, Center for Bioethics, Harvard Medical School,
Boston, MA, USA

Institute for Professionalism and Ethical Practice, Boston, MA, USA

Critical Care Medicine, Boston Children's Hospital, Boston, MA, USA

C.M. Ulrich
Lillian S. Brunner Endowed Chair, University of Pennsylvania School of Nursing,
Philadelphia, PA, USA

Department of Medical Ethics and Health Policy, Perelman School of Medicine,
University of Pennsylvania School of Medicine, Philadelphia, PA, USA

© Springer International Publishing AG 2018   159
C.M. Ulrich, C. Grady (eds.), *Moral Distress in the Health Professions*,
https://doi.org/10.1007/978-3-319-64626-8_9

Everyone values quality and compassionate healthcare, especially when they need it for themselves or someone they care about. Recent debates have highlighted the complexity and multifaceted nature of healthcare, both in the United States and abroad. Patients expect and deserve knowledgeable, responsible, and caring healthcare providers. Yet, providing healthcare can be hard work, both physically and emotionally. The daily work of healthcare providers can be rewarding and inspiring, and simultaneously stressful and very sad. Unfortunately and not infrequently, in addition to other stressors, healthcare providers experience moral distress when they feel unable to do what they think is right, or feel bad about or regret their involvement in a morally undesirable situation. We, the editors, compiled this book because we believe that moral distress is a phenomenon experienced by healthcare providers of many disciplines, for many reasons, and in healthcare settings across the globe. We also believe that healthcare providers have many moral strengths and regularly find ways to successfully navigate ethical challenges, learn from their experiences, and flourish.

We set out to ask a number of thought leaders to help us think about two important questions related to moral distress (1. What do you think is the most significant or important reason that healthcare professionals might feel trapped and unable to do what they think is right? 2. What does moral success look like to you?--)—reasons for moral distress and successes in spite of challenges. Several thoughtful people responded to our queries and we highlight their thoughts in this concluding chapter. We also elaborate on their contributions with observations from the literature and our own thoughts. Overall, more normative and empirical bioethics work is needed to find ways to minimize moral distress and to help clinicians who are struggling with it. All of our thought leaders underscore concerns related to feeling powerless and voiceless, lack of supportive environments (including leadership concerns), and feeling trapped by requests for aggressive procedures. Our thought leaders also share innovative as well as provocative ways to turn around a potentially distressing situation and ultimately make it a "moral success." We share and expand on these responses below.

9.1   Significant Reasons for Moral Distress

In a prominent New York Times article in 2009 entitled "Why Doctors and Nurses Cannot Do the Right Thing," Pauline Chen (surgeon and author) writes that "Doctors and nurses "feel trapped," … by the competing demands of administrators, insurance companies, lawyers, patients' families, and even one another. And they are forced to compromise on what they believe is right for patients… (moral distress)." Dr. Chen shares her personal journey and lived experience as a surgeon in her book "Final Exam." Although she doesn't directly use the term "moral distress" she describes many reasons for the moral distress and moral residue that we see in day-to-day clinical practice: for example, the "turfing" of difficult patients to others to avoid sensitive conversations, the "intoxicating power of treatment," and the stress of caring for the dying and the grieving.

Ann B. Hamric (Professor Emerita, School of Nursing, Virginia Commonwealth University and Bioethics Nursing Scholar), well-known for her extensive research and scholarship on moral distress, writes: "As research demonstrates, there are many important root causes of moral distress—this reality is one of the reasons that moral distress is so challenging to address. So I resist identifying one "most significant or important reason." With that said, I would point to work environments that do not prioritize and support safeguarding the moral integrity of their healthcare professionals. In these environments, some of which claim to be "healthy work environments," there is little explicit discussion of the *ethical* dimensions of practice, interprofessional collaboration is given lip service rather than teams undertaking the hard effort to make true collaboration a practicing reality, and administrative leadership does not safeguard professionals' moral integrity, either by devaluing conflicting perspectives, ignoring/avoiding morally distressing issues or refusing to make difficult decisions. The fact that respondents in our research continue to make comments such as "I am afraid to speak up" is indicative of this failure to achieve an ethical practice environment and the broad repercussions it has for perpetuating patient-, unit-, and system-level moral distress."

Arthur Caplan (Professor and Director of the Division of Medical Ethics at New York University, and a well- known bioethicist) explains the reasons for distress in different terms, but ultimately indicts powerlessness in the face of lack of support in the work environment as well. In response to our question, he wrote that the most important reason for moral distress is feeling powerless, and the anger and indifference that follow when not knowing what to do. He says: "In my experience the single most important source of moral distress is the feeling of powerlessness. Health care providers often perceive a problem but have no idea where to go with their concerns or worse feel it is a sign of weakness or inexperience if they raise an ethical concern. They often see the same problem arising again and again but don't know where to turn to change the system or get at the underlying cause. So anger often replaces concern and then that is too often followed by accommodation and then indifference."

Nancy Berlinger (Bioethics Research Scholar at The Hastings Center) also notes that moral distress arises from a feeling of powerlessness, and emphasizes that feeling powerless is often based on inadequate support within healthcare systems. She says: "The experience of moral distress is an individual (or, in some cases, group) response to a structural problem of complex organizations, notably healthcare systems. The person (or group) experiencing moral distress perceives two things. One is a powerful moral intuition concerning right or wrong action, the other is a powerful feeling of distress associated with perceived powerlessness to remedy the situation. Thus, the morally distressed person feels forced to do what is wrong, or prevented from doing what is right. It's important to keep in mind that either or both perceptions could be misperceptions. The person experiencing a gut-level intuition that something is "not right" could be wrong on the facts, or could be ascribing a general feeling of distress in the face of suffering to the narrower category of moral distress. The person who perceives herself to be powerless could be wrong concerning her ability to take action. With these caveats in mind—to avoid conflating a range of valid but distinct professional and workplace concerns into moral distress—one of the most significant reasons healthcare professionals feel that "trapped" feeling is because an issue that has been unresolved at some other level of a system has been pushed onto them. So, when a professional feels "trapped" into participating in an intervention with little potential to benefit a person nearing the end of life, the unresolved issue may be one of training and of structural support: how well or poorly do our systems of professional training and institutional support prepare and help professionals, patients, surrogate decision-makers, and other loved ones to communicate with one another about the benefits and burdens of interventions when a person is nearing the end of life? Or, when a professional feels "trapped" by limited resources in the care of an uninsured patient, the unresolved issue is likely in the realm of organizational or public policy. In each case, another level of the system has failed, or avoided seeing, a foreseeable problem that will fall on a bedside clinician or unit-level manager to try to resolve, and yet, these person's hands are not on the levers that can shift a policy in a different direction. Exhortations directed solely at this professional—whether to "be a good advocate," or to show "moral courage," or to be "resilient"—cannot resolve a system-level problem, and are ethically unsound if they overlook the consequences *for patients* of systemic problems that produce moral distress in professionals."

Shaké Ketefian (Professor Emerita, University of Michigan, and a pioneer in nursing ethics) also cites lack of power and lack of support as important sources of moral distress. She described her research and development of *Judgments about Nursing Decisions*—JAND (see Box 9.1) as a tool for gauging discrepancies between what nurses thought they ought to do in certain situations, and what they thought would actually be done. She identified several reasons derived from JAND data to explain the discrepancies that she found, and notes that these and others have been mentioned in the literature, including: "The environment where they worked was not supportive of ethical practice; They were overruled by others; Nursing leadership does not support nurses when individual nurses stand up for patients or their own principles; Nurses were afraid of losing their jobs; Collegiality and mutual support between and among health professionals were lacking." She then goes on to say:

Yet, I cannot help but think that at least in some cases nurses engage in "self-censorship," which is to say that they tend to anticipate that if they engaged in an action or took a position, it might antagonize some parties that they perceive to be powerful, or have power over them, and as a result, withhold from taking a position or action that would be in accordance with their professional views on behalf of patients.

There are of course other factors as well, to which this entire volume is devoted; one major issue in my view is that nurses as a group or individually do not seem to have power in most systems. This is rather strange in view of the fact that repeated Gallup polls have shown that the public trusts nurses above all other professional groups, considering them honest and as having the highest ethical standards. For some reason nurses have not leveraged this big plus into power for themselves on behalf of patients within healthcare organizations. On the contrary, some institutions are re-naming their health systems to include name of institution followed by 'medicine,' such as *Michigan Medicine*, thus giving recognition only to medical providers in their health systems, and explicitly denying the contributions of other professional groups to the healthcare of patients. Furthermore, and sadly, neither has this national trend been met by an outcry, loud or otherwise, from the various health professional groups that contribute to health and patient care, whether it be nursing, pharmacy, physiotherapy, social work, to name just a few of the professions.

Box 9.1 Shaké Ketefian on the Development of *Judgments about Nursing Decisions*

"In the late 1970s I embarked upon research on nursing decisions/judgments on ethical matters nurses encounter in their daily work. The first task I faced was one of measurement: there was no existing tool to measure the ethical quality of nurses' decisions. The *Moral Judgment Interview* (MJI) that Professor L. Kohlberg of Harvard University had developed focused on moral development or reasoning, also referred to as moral judgment development, which deals with the way people reason about moral choice. Same with the *Defining Issues Test* (DIT) by Professor J. Rest, of the University of Minnesota, which was a paper and pencil test, and a more efficient way to get at moral reasoning than the MJI. The research question I wanted to pose was: what is the relationship between moral reasoning and judgments made in real life ethical dilemmas, presented as simulated scenarios? In other words, if a person reasons well and at high levels, which in the language of Kohlberg would be at post-conventional levels, does it follow that they will necessarily make ethical decisions? Strictly speaking one does not follow from the other, although there is an implicit expectation that at high levels of post-conventional reasoning, that take into account justice considerations, the person's decision would be ethically sound. At that time we were entering a new era when a few of us were beginning to conduct empirical research on ethical matters; until that point, research in the ethical practice domain was done via philosophical discourse.

I was not sure what the shape of my new instrument might be, so I had to start with an open mind. As a first step I convened a group of seasoned nurses in graduate school to talk about their experiences in their practice on matters ethical, and to discuss ethical conflict situations written by other nurses drawn from their own practice. I asked my group the question: what are possible actions a nurse might take in the dilemmas presented, whether the actions were ethical or not.

This process led to a large list of actions for each of over 10 dilemmas nurses were likely to encounter. At the same time, the insight emerged that there may be a distinction between what a nurse should do/considers appropriate ethical action, and what s/he is likely to do in a real-life situation. This would have major implications for how the instrument might be structured. Another issue to contend with during this process had to do with the *standard* to be used to assess whether a given decision/action was ethical or not. This issue was resolved by using the ANA code of ethics of that era as the standard.

In addition to dealing with the quality of nursing actions the structure of the ethical practice/behavior instrument had to somehow capture the distinction between actions *that the nurse believes should or should not be taken* in a given situation presenting ethical conflict, distinct from *what actions s/he believes are in fact likely to be taken* in her/his work environment. This thinking led to building two columns to which respondents were asked to answer: indicate whether they believe each action should be taken or not (column A), and then, whether in their unit the action is realistically likely to be taken by the nurses (column B). In this format, column A represented what the respondents knew to be ethical action, and column B represented what action the respondent thought her/his colleagues were likely to take. However, in no place did we directly ask what the respondent would do; the reason for this was to remove the issue of subjectivity from interfering with responses. Most human beings are not able to assess their own actions and behavior with a credible level of objectivity.

In analyzing the instrument we found that the majority of all samples indicated that they understood the nature of the conflict and identified the correct ethical actions in column A. However, many of those who knew the ethical actions did not carry through, and chose that non-ethical actions would be taken by nurses in their units (column B). This significant level of discrepancy has persisted both in U.S. samples and other international samples as well. This situation creates and fits the precise definition of moral distress that Jameton (1984) described a few years later in his book."

Sheila Davis (Chief Nursing Officer and the Chief of Ebola Response at Partners In Health) describes a significant source of moral distress as the powerlessness that healthcare providers undergo when encountering patients who cannot get the healthcare they need without recourse: "Working in resource poor settings in the United States and other sites globally, the stark reality of the lack of accessible and quality healthcare for the majority of people in the world is distressing for all who see it. Most healthcare professionals working on humanitarian relief efforts post-disaster expect to see harsh conditions and their effort is usually time-limited and focused on life saving surgeries, assistance with basic necessities such as food/water and other acute disaster related efforts. Although immediate post-disaster is difficult and can cause moral distress due to the sheer magnitude of suffering and aftermath of the acute disaster, more troubling to me is the ongoing chronic disaster of poverty and lack of healthcare." Although she highlights the particular distress felt by humanitarian workers, Sheila recognizes the plight of healthcare workers in general during a public health emergency. Sheila shares the trauma and distress that healthcare providers experienced caring for Ebola patients under extremely difficult circumstances. (see Box 9.2). Connie Ulrich poignantly calls attention to the moral distress felt by African healthcare providers during the Ebola crisis, noting that this group of healthcare providers rose to the challenge of caring for extremely ill patients with "…resolve, resiliency, and commitment to patient care in the face of extreme adversity" [1]. Finding ways to help mitigate or address moral distress in such circumstances is a monumental and important task.

> **Box 9.2 Sheila Davis—Moral Distress Facing Health Care Volunteers Treating Ebola Patients in Sierra Leone**
> "The moral distress experienced during the Ebola crisis in West Africa was extraordinary. Healthcare providers were not able to provide the skin-to-skin therapeutic touch to human beings who were suffering and dying and this was very difficult. Providing the simplest care–intravenous fluids, oral rehydration solution, and the keeping patients as clean as possible wearing a face shield, hood, tall rubber boots, double gloves and full plastic body suit (personal protective equipment or PPE) was technically and physically challenging. More difficult was the psychological stress of fearing harm to oneself or co-workers and the moral dilemmas that were faced when our own personal safety had to be weighed in relation to care we could provide our patients. The time in the Ebola Treatment Unit (ETU) was closely monitored to be no more than one hour to prevent staff from becoming dehydrated in the 90+ degree heat that could put themselves or others at risk. This policy was often difficult to enforce because the healthcare workers did not want to leave the patients and became angry when they had to leave the ETU.
>
> While working in Sierra Leone during the 2014–2015 Ebola crisis, it soon became apparent that although we could treat many people with Ebola and successfully discharge them back into the community, many were too weak to return home. The elation of a successful discharge from our ETU dampened with the harsh reality of the near-by district hospital that was the only option for patients' post-Ebola care. The hospital was poorly staffed, had few medications, no electricity and many other challenges. Our strategy quickly changed and we began detouring some of our Ebola clinicians to the hospital to provide care. Many of the short-term clinicians we employed to work with us during our Ebola response were new to global health and thought the lack of staff, medication, and poor infrastructure including water and electricity were a result of the emergency efforts against Ebola. The few national staff that were still working at the hospital quickly set the record straight that this was what the hospital was like pre-Ebola and this was the reality of healthcare in Sierra Leone.
>
> We split our staff between the ETU and the hospital and assumed the hospital rotation would be a "break" from the stress of working in the horrible conditions of the ETU. We were wrong as many of our staff found the hospital duties much more stressful, and burnout and disillusionment with the work was more evident. Our staff expressed moral outrage that the state of the hospital was the "status quo" and became angry that we could not do more to help. They had come with the realization that some people with Ebola would

tragically die, but they were not prepared for infants, children and adults dying of completely preventable and treatable conditions and diseases. Sadly, many of us who are permanent global health staff that work in very poor countries around the world all of the time are socialized for scarcity and the situation was less shocking. I still believe we were approaching our care with compassion, but the desensitization to human suffering can still be an unfortunate, and unacceptable, hazard of global health. Faced with abject suffering and few resources to change the circumstances during acute or chronic disasters, heath care workers experience moral distress that can have devastating personal and professional consequences when one is unable to reconcile their lived experience with those who are familiar with the unique context."

Robert Truog (a pediatric intensivist and Director of the Center for Bioethics at Harvard Medical School) eloquently describes how healthcare providers, especially in a critical care setting, might feel trapped by having to perform cardiopulmonary resuscitation (CPR) on a patient who is clearly and imminently dying. In these cases, the healthcare provider might think that CPR is inappropriate despite the family or surrogate decision maker's insistence.

9.2   Examples of Moral Success

Arthur Caplan describes moral success as action: "…moral success represents not just perceiving a problem but finding ways to resolve it that are actionable and practical, rather than idealistic and grandiose." He then goes on to give an example: "I have enjoyed a number of moral successes in my own work from persuading hospitals' home care programs to institute mandatory flu vaccination requirements on the grounds that healthcare providers have a duty to prevent harm, protect the vulnerable and to put patient interests first, to helping a resident find support and raise money to send a severely injured patient from Bellevue Hospital to return to his native Poland both to see family and perhaps to die among them. The resident took the initiative, got involved in every detail and worked out a solution. He asked various people for help and I was thrilled and proud to help him. He was a good and ethical doctor to begin with. He was a better one for what he did to help his patient."

Ann Hamric shares a story of moral success via collective action when practicing as a clinical nurse specialist on a neurosurgical/orthopedic surgical unit with complex patients and a nursing staff shortage. "We had a protracted nursing shortage which we tried to manage with many different strategies, such as aggressive recruitment, use of travelers, floating staff from other areas, and using less skilled staff. Over the course of months, none of the strategies were effective and patient care began to deteriorate: a unit that had prided itself on the quality of its nursing care began to have patients with complications due to our inability to give all the patients the care they required. I documented these concerns repeatedly to administration, but while the Director of Nursing was sympathetic, nothing changed. I even documented the particular patient complications that we knew were attributable to inadequate nursing care. We requested that beds be closed to allow for a more appropriate patient: nurse ratio, but the Chief of Neurosurgery was incensed by this idea. Finally, the staff said the only thing they could think to do was go on strike (this was not a unionized hospital), and they asked for my support. This threat precipitated a meeting with the staff, the CEO, the Chair of Neurosurgery and the Director of Nursing. The staff clearly articulated their moral distress (though we did not have that term then!) over the repeated compromises they had to make in providing care to our complex and mostly bedridden patients, and their unwillingness to continue without some relief. Within days, the administrative leaders agreed to close four beds, as we requested, and supported our desire to have adequate staffing ratios to deliver the quality of care we knew how to deliver. The beds stayed closed until we got our staffing numbers up. That was a moral success, though it was a long time coming."

Bob Truog explains why feeling trapped into providing "inappropriate" CPR can cause moral distress, and then suggests that by adopting the perspective of the family or surrogate, and a feasible, yet perhaps controversial, solution, one can find a "success" that mitigates moral distress (see Box 9.3).

Box 9.3 Robert Truog—An Innovative Way to Mitigate Moral Distress from Inappropriate CPR

Moral distress is emerging as a critically important issue for the psychological and physical well-being of clinicians, and in particular those working in critical care environments. In my experience, one of the most high-risk situations is when family members insist on cardiopulmonary resuscitation (CPR) for patients who are imminently dying.

Doctors and nurses often experience this situation as one of senseless and inappropriate violence and brutality. In performing chest compressions, we are subjecting the patient to pain and suffering at precisely the time when we believe the focus should be on creating a comfortable and peaceful environment for everyone involved.

I've experienced these feelings myself, and I don't want to minimize them in any way. But I've also found it helpful at times to reflect on what the experience might be like from the perspective of the patient and family themselves. The experience of the healthcare system for those who come from underserved backgrounds is often one of having been repeatedly denied access to essential medical services, and the perception that they are now being denied a potentially life-saving intervention (that is, CPR) can be seen as the final insult in a long series of past injustices. Or for those who have family roots in the developing world, the thought of having to tell the relatives "back home" that at the end of their loved-one's life they agreed to "give up" and forego CPR may be intolerable, particularly when they know that their extended family overseas has no ability to understand the limitations of a high-tech healthcare system that they have never experienced. And most importantly, in many such cases we know that the patient has herself insisted on receiving CPR, even if only to help assuage the grief of her family and friends.

In situations like these, I have sometimes found that a brief "code" can be conducted in a way that allows the family to feel that "everything" has been done and with little risk of pain and suffering for the patient, particularly in situations where the patient is mechanically ventilated and generous amounts of sedation and analgesia can be administered beforehand.

This approach is, admittedly, controversial. So called "slow codes" have been criticized and reviled for decades. But if we believe we are treating not only the patient, but also the family who will live with their memories of this event for a long time to come, I think that sometimes this may be justified. And instead of experiencing the moral distress that comes with telling others "I feel horrible today because I was forced to assault a dying patient at the demands of an unreasonable family," I may be able to say, "Today I did something that I would never want for myself or anyone I loved, but I did what I thought was right at the time for this particular patient and family."

Shaké Ketefian describes education of Doctors of Nursing Practice as key to developing the confidence and skills that nurses need for moral success. She asks "How do nurses gain power and concurrent "voice" that goes with it in patient care decisions? By power I do not refer to raw political power over anyone. Rather, the power I have in mind is the kind that comes from the exercise of unique nursing knowledge and expertise nurses bring to patient care, and the self-confidence that comes with the acquisition of such expertise. How can this be done?" Her answer is higher education, and she writes:

Nursing has taken the major step of developing a Doctor of Nursing Practice (DNP) degree in the past decade, which has spread quickly nationwide. With this professional doctorate, the graduates will have an education, knowledge base and expertise commensurate with that of other health professionals, which will give them the professional qualifications and the confidence to offer views and perspectives that have equal weight along with the views of others.

In order for needed changes to come about, the DNP graduates need to populate the practice arena... Currently, the majority of practicing nurses are associate degree, diploma, or baccalaureate degree (entry level professional preparation) graduates... Educational preparation at these levels is not sufficient to provide in-depth understanding of the complex, multidimensional problems patients present, or to deal with professional issues or health policy, nor would they have had the opportunity to develop intellectual agility and skills, or the necessary self-confidence to weigh-in during discourse with other professionals and be viewed with credibility. But I believe DNP graduates would be viewed as credible peers…. There is an urgent need to address the development of institutional policies and standards, and promote interprofessional dialogue and understanding to bring about supportive and ethical work environments to enable nurses to provide optimal patient care and bring their voices to bear in decisions concerning patients. In all of these activities a critical mass of DNP-prepared nurses can have major impact.

Ann Hamric and Elizabeth Epstein describe a pathway to promoting moral success through a moral distress consultation service. They incorporated a moral distress consultation service into the more traditional ethics consultation service that is often part of healthcare organizations. In an evaluation of their consultation service experience, they report that moral distress consultations were conducted on 25 different units, including intensive care, and other acute and outpatient areas, they identified more than 30 different root causes of moral distress from 56 consults (including, e.g., inadequate team communication, lack of continuity in care, unclear treatment goals, futility concerns, abusive families, and more), and they conclude that a system-wide approach is warranted. These authors found that the opportunity to dialogue and confront morally distressing issues with interdisciplinary colleagues not only gave some legitimacy to the experience of moral distress, but also led to an empowered voice (particularly for nursing staff), an increased sense of collaboration, and engagement for organizational or unit change [2].

Nancy Berlinger describes moral distress as a "…collective action problem. It is produced by a system, it is experienced by individuals or groups on the lower, receiving, end of hierarchies, and it can't be resolved by an individual or a lower-status group, so its resolution, including the analysis of upstream problems, depends on more-powerful individuals or groups taking an interest on behalf of the system and those it includes and those it serves. So moral success—if we're taking this term to be the opposite of or antidote to moral distress—must be more than a moral distress-free day at work, or being "ethical." It must also have a systemic dimension, and must involve some effort to get at factors that produce moral distress. One example of moral success would be when morally distressed professionals agree that simply feeling bad, even tortured, does not, itself, improve conditions for their patients, and further agree to take the first steps toward collective action: airing their perspectives, getting at what, exactly, triggers that "trapped" feeling in some cases, but perhaps not in others, and identifying opportunities for further action. This takes time, space, a skilled facilitator, and a goal beyond venting or mutual support. Once it's clear whether there is an actionable problem, separate from the feelings associated with the problem, the routes to action may be clearer. For example, if professionals working in the same unit recognize that the dual perceptions associated with moral distress—wrongness and helplessness—are triggered by cases in which a patient cannot gain access to a medically appropriate service available to other patients, for reasons such as undocumented status, homelessness, or dual diagnosis, the next step to alleviating moral distress would be investigating whether or not there are ways for these patients to receive the care that they need. This is likely to be most effective if done collectively, and by engaging the next level of the hierarchy, rather than by parallel advocacy efforts on behalf of individual patients; this may feel like success, even moral success, but leaves the basic problem—can access be expanded?—untouched. This next level of engagement will take more time, space, facilitation, and the articulation of an achievable goal, one that is likely to require the spending of political capital within a system: can system leaders be convinced to invest in services that will compensate for a barrier to care, or to advocate for public policy reform to expand access to include an excluded group? If moral success is to be a meaningful concept in patient care, the success should be experienced on behalf of patients, with an easing of professional distress as a result, rather than the main goal.

Cynda Rushton (a distinguished nurse bioethicist at Johns Hopkins University) has championed a concept of "moral resilience." Rushton and Alisa Carse note that "It is crucial that we find ways to empower clinicians in heeding this call-to support clinicians' moral agency and voice, foster their moral resilience, and facilitate their ability to contribute to needed reform within the organizations and systems in which they work" [3]. Rushton also notes the need for additional conceptual work to help refine the meaning of moral resilience and how to find ways to employ resilience in mitigating the negative effects of moral distress. [4] Sheryl Sandberg and Adam Grant discuss a concept of "collective resilience," in their book entitled "Option B: Facing Adversity, Building Resilience, and Finding Joy." [5] Although their focus is not on healthcare or healthcare provider moral distress, they note that "by helping people cope with difficult circumstances and then taking action to alter those circumstances, collective resilience can foster real social change." ([5], p. 135). In their view, "collective resilience requires more than just shared hope—it is also fueled by shared experiences, shared narratives, and shared power" (p 130). Moral distress has created a certain sense of "shared identity" among healthcare clinicians, and perhaps shared experiences, narratives, and power could promote collective resilience.

9.3   Summary

Throughout this book, our colleagues from multiple healthcare disciplines share their thoughts and intimate stories on the experience and impact of moral distress in their work lives and the ways in which a particular patient case or situation, along with the work climate or ethos of a particular institution or setting, has challenged their moral fabric. Sharing narrative stories is one way to air both the positive and negative experiences that a community of clinicians might face in their clinical practice; we hope that the diverse stories on moral distress shared in our book offer one path forward in communicating about the pervasive nature of moral distress and building stronger interdisciplinary communities for the betterment of patient care.

We are still left with many unanswered questions about a complex phenomenon and a need for rigorously tested strategies useful across disciplines, settings, and geographical boundaries, to prevent, mitigate, or treat moral distress. Moral distress is a phenomenon that continues to require both normative and empirical analyses as clinicians strive to fulfill their moral responsibilities to patients, families, and communities. Strategies are needed that help clinicians to forestall, prevent, or overcome the sense of powerlessness that is so often equated with moral distress. Also necessary is identification and evaluation of strategies to create supportive workplaces and systems that are responsive to required change and promote moral success. Further, we need to examine the characteristics of successful teams and organizations and the qualities that are most conducive to moral success while engaging with ethical issues within healthcare systems.

References

1. Ulrich C. Ebola is causing moral distress among African healthcare workers. BMJ. 2014;349:g6672.
2. Hamric A, Epstein E. A health system-wide moral distress consultation service: development and evaluation. HEC Forum. 2017. https://doi.org/10.1007/s10730-016-9315-y.
3. Carse A, Rushton C. Harnessing the promise of moral distress: a call for re-orientation. J Clin Ethics. 2017;28(1):15–29.
4. Young PD, Rushton CH. A concept analysis of moral resilience. Nurs Outlook. 2017. https://doi.org/10.1016/j.outlook.2017.03.009.
5. Sandberg S, Grant A. Option B: facing adversity, building resilience, and finding joy. New York: Knopf/Borzoi Books; 2017.